CAMBRIDGE STUDIES IN ADVANCED
MATHEMATICS 115

EDITORIAL BOARD

Mathematical Tools for One-Dimensional Dynamics

Originating with the pioneering works of P. Fatou and G. Julia, the subject
of complex dynamics has seen great advances in recent years. Complex
dynamical systems often exhibit rich, chaotic behavior, which yields
attractive computer generated pictures, for example the Mandelbrot and
Julia sets, which have done much to renew interest in the subject.

In this self-contained book, the major mathematical tools necessary for
the study of complex dynamics at an advanced level are discussed. Complete
proofs of some of these tools are presented; some, such as the Bers–Royden
theorem on holomorphic motions, appear for the first time in book format.
Riemann surfaces and Teichmüller theory are considered in an appendix.
Detailing the latest research, the book will appeal to graduate students and
researchers working in dynamical systems and related fields. Carefully
chosen exercises aid understanding and provide a glimpse of further
developments in real and complex one-dimensional dynamics.

CAMBRIDGE STUDIES IN ADVANCED MATHEMATICS

Editorial Board:

B. Bollobas, W. Fulton, A. Katok, F. Kirwan, P. Sarnak, B. Simon, B. Totaro

All the titles listed below can be obtained from good booksellers or from Cambridge University Press. For a complete series listing visit:
http://www.cambridge.org/series/sSeries.asp?code=CSAM

Mathematical Tools for One-Dimensional Dynamics

EDSON DE FARIA
IME-USP, São Paulo

WELINGTON DE MELO
IMPA, Rio de Janeiro

CAMBRIDGE
UNIVERSITY PRESS

Shaftesbury Road, Cambridge CB2 8EA, United Kingdom

One Liberty Plaza, 20th Floor, New York, NY 10006, USA

477 Williamstown Road, Port Melbourne, VIC 3207, Australia

314–321, 3rd Floor, Plot 3, Splendor Forum, Jasola District Centre, New Delhi – 110025, India

103 Penang Road, #05–06/07, Visioncrest Commercial, Singapore 238467

Cambridge University Press is part of Cambridge University Press & Assessment, a department of the University of Cambridge.

We share the University's mission to contribute to society through the pursuit of education, learning and research at the highest international levels of excellence.

www.cambridge.org
Information on this title: www.cambridge.org/9780521888615

First published 2008

A catalogue record for this publication is available from the British Library

Library of Congress Cataloging-in-Publication data
Faria, Edson de.
Mathematical tools for one-dimensional dynamics / Edson de Faria and Welington de Melo.
p. cm.
Includes bibliographical references and index.
ISBN 978-0-521-88861-5 (hardback)
1. Dynamics. 2. Teichmüller spaces. 3. Riemann surfaces.
4. Holomorphic functions. I. Melo, Welington de. II. Title.
QA845.F37 2008
531´.11–dc22   2008021976

ISBN   978-0-521-88861-5   Hardback

Contents

Preface

Our main goal in this book is to introduce the reader to some of the most useful tools of modern one-dimensional dynamics. We do not aim at being comprehensive but prefer instead to focus our attention on certain key tools. We believe that the topics covered here are representative of the depth and beauty of the ideas in the subject. For each tool presented in the book, we have selected at least one non-trivial dynamical application to go with it.

Almost all the topics discussed in the text have their source in complex function theory and the related areas of hyperbolic geometry, quasiconformal mappings and Teichmüller theory. This is true even of certain tools, such as the distortion of cross-ratios, that are applied to problems in *real* one-dimensional dynamics. The main tools include three deep theorems: the *uniformization theorem* (for domains in the Riemann sphere), the *measurable Riemann mapping theorem* and the *Bers–Royden theorem* on holomorphic motions. These are presented, with complete proofs, in Chapters 3, 4 and 5 respectively.

The present book originated in a set of notes for a short course we taught at the 23rd Brazilian Mathematics Colloquium (IMPA, 2001). We have benefited from useful criticism of the original notes from friends and colleagues, especially André de Carvalho, who read through them and found several inaccuracies. We are also grateful to two anonymous referees for their perspicacious remarks and suggestions.

The drawings of Julia sets and the Mandelbrot set found in this book were made with the help of computer programs written by C. Mullen (available through his homepage, at www.math.harvard.edu/~ctm). We wish to thank also Dayse H. Pastore for her help with the figures for the original colloquium notes, some of which appear in the present book.

Finally, it is quite an honor to see our book published by Cambridge University Press, and in such a prestigious series. We are grateful to David Tranah and Peter Thompson for this opportunity and for their patient and highly professional support.

Edson de Faria and Welington de Melo
São Paulo and Rio de Janeiro
January 2008

1
Introduction

It is fair to say that the subject known today as complex dynamics – the study of iterations of analytic functions – originated in the pioneering works of P. Fatou and G. Julia early in the twentieth century (see the references [Fat] and [Ju]). In possession of what was then a new tool, Montel's theorem on normal families, Fatou and Julia each investigated the iteration of rational maps of the Riemann sphere and found that these dynamical systems had an extremely rich orbit structure. They observed that each rational map produced a dichotomy of behavior for points on the Riemann sphere. Some points – constituting a totally invariant open set known today as the *Fatou set* – showed an essentially dissipative or wandering character under iteration by the map. The remaining points formed a totally invariant compact set, today called the *Julia set*. The dynamics of a rational map on its Julia set showed a very complicated recurrent behavior, with transitive orbits and a dense subset of periodic points. Since the Julia set seemed so difficult to analyse, Fatou turned his attention to its complement (the Fatou set). The components of the Fatou set are mapped to other components, and Fatou observed that these seemed to eventually to fall into a periodic cycle of components. Unable to prove this fact, but able to verify it for many examples, Fatou nevertheless conjectured that *rational maps have no wandering domains*. He also analysed the periodic components and was essentially able to classify them into finitely many types.

It soon became apparent that even the *local* dynamics of an analytic map was not well understood. It was not always possible to linearize the dynamics of a map near a fixed or periodic point, and remarkable examples to that effect were discovered by H. Cremer. In the succeeding decades researchers in the subject turned to this linearization problem,

1

and the more global aspects of the dynamics of rational maps were all
but forgotten for about half a century.

With the arrival of fast computers and the first pictures of the Mandelbrot set, interest in the subject began to be revived. People could
now draw computer pictures of Julia sets that were not only of great
beauty but also inspired new conjectures. But the real revolution in the
subject came with the work of D. Sullivan in the early 1980s. He was
the first to realize that the Fatou–Julia theory was strongly linked to the
theory of Kleinian groups, and he established a *dictionary* between the
two theories. Borrowing a fundamental technique first used by Ahlfors
in the theory of Kleinian groups, Sullivan proved Fatou's long-standing
conjecture on wandering domains. With this theorem Sullivan started a
new era in the theory of iterations of rational functions.

Our goal in this book is to present some of the main tools that are
relevant to these developments (and to other more recent ones). Our
efforts are concentrated on the exposition of only a few tools. We tried
to select at least one interesting dynamical application for each tool presented, but it was not possible to be very systematic. Many interesting
techniques had to be omitted, as well as many of the more interesting contemporary applications. There are a number of superbly written
texts in complex dynamics with a more systematic exposition of theory;
we strongly recommend [B2], [CG], [Mi1], [MNTU], as well as the more
specialized [McM1, McM2].

The remainder of this introduction is devoted to a more careful expla-
nation of the basic concepts involved in the above discussion and also to
a brief description of the contents of the book.

Let $\widehat{\mathbb{C}} = \mathbb{C} \cup \{\infty\}$ be the Riemann sphere, $\text{Poly}_d(\mathbb{C})$ be the space of
polynomials of degree d and $\text{Rat}_d(\widehat{\mathbb{C}})$ be the space of rational functions
of degree d, $d \geq 2$. An element $f \in \text{Rat}_d(\widehat{\mathbb{C}})$ is the quotient of two
polynomials of degree $\leq d$. If the derivative of f at p vanishes or,
equivalently, if f is not locally one to one in any small neighborhood
of p, we say that p is a *critical* point of f and its image $f(p)$ is a
critical value of f. When f is a polynomial of degree d, the point
∞ is always a critical point of multiplicity $d - 1$ (the polynomial f
is d to one in a neighborhood of ∞). The number of points in the
pre-image of a point that is not a critical value is constant and equal
to the degree of the rational map f. The sum of the multiplicities of
critical points of a rational map of degree d is equal to $2d - 2$. In
particular, a polynomial of degree d has $d - 1$ finite critical points. The
iterates of f are the rational maps $f^1 = f$, $f^n = f \circ f^{n-1}$. The *forward*

orbit of a point p is the subset $O^+(p) = \{f^n(p), n \geq 0\}$, its *backward orbit* is $O^-(p) = \{w \in \mathbb{C}; f^n(w) = p, n \geq 0\}$ and its *grand orbit* is $O(p) = \{w \in \widehat{\mathbb{C}}; f^n(w) = f^m(p), m, n \geq 0\}$.

Two maps f, g are *topologically conjugate* if there is a homeomorphism $h : \widehat{\mathbb{C}} \to \widehat{\mathbb{C}}$ such that $h \circ f = g \circ h$. It follows that $h \circ f^n = g^n \circ h$ for all n and hence that the *conjugacy* h maps orbits of f into orbits of g. Since h is continuous, it preserves the asymptotic behavior of the orbits. A rational map f is *structurally stable* if there is a neighborhood of f in the space of rational maps of the same degree such that each map in this neighborhood is topologically conjugate to f.

We will consider also some special analytic families of rational maps. By such a family we mean an analytic map $F \colon \Lambda \times \widehat{\mathbb{C}} \to \widehat{\mathbb{C}}$, where Λ is an open set of parameters in some complex Banach space, such that $F_\lambda :$ $z \mapsto F(\lambda, z)$ is a rational map for each $\lambda \in \Lambda$. The space of polynomials of degree d is an example of an analytic family of rational maps. We will also consider the notion of structural stability with respect to such a family: F_λ is structurally stable, with respect to that family, if there is a neighborhood of λ in the parameter space Λ such that, for μ in this neighborhood, F_μ is topologically conjugate to F_λ. The complement of the stable parameter values is called the *bifurcation set* of the family. It is clearly a closed subset of the parameter space Λ.

Given a rational map f, the phase space $\widehat{\mathbb{C}}$ decomposes into the disjoint union of two totally invariant subsets, the *Fatou* set $F(f)$ and the *Julia* set $J(f)$. A point z belongs to the Fatou set if there exists a neighborhood V of z such that the restrictions of all iterates f^n to this neighborhood form an equicontinuous family of functions that is, by the Arzelá–Ascoli theorem, a pre-compact family in the topology of uniform convergence on compact subsets. Therefore the Fatou set is an open set where the dynamics is simple. The Julia set, its complement, is a compact subset of the Riemann sphere. The topological and dynamical structure of these sets was the main object of study of Fatou [Fat], Julia [Ju] and others, using compactness results for families of holomorphic functions such as Montel's theorem, mentioned above, and Koebe's distortion theorem. These tools will be discussed in Chapter 3.

As mentioned earlier, a complete understanding of the structure of the Fatou set for any rational map had to wait until the 1980s, when Sullivan brought to the subject the theory of deformations of conformal structures. We will discuss this theory in Chapter 4. Sullivan proved the *no-wandering-domains* theorem [Su], which states that each connected component of the Fatou set is eventually mapped into a periodic component and that the number of periodic cycles of components is bounded. See section 4.6 for the proof of Sullivan's theorem.

The two main results of the deformation theory of conformal structures are the Ahlfors–Bers theorem, which will be discussed in section 4.4, and the theorem on the extensions and quasiconformality of holomorphic motions, which will be discussed in Chapter 5. Using these two important tools, it is proved in [MSS] and in [McS] that the set of stable parameter values is dense in any analytic family of rational maps. In particular the set of structurally stable rational maps is open and dense. See section 5.4 for a proof of this fundamental structural stability result. However, the bifurcation set is also a large and intricate set. In fact, M. Rees proved in [Re] that, in the space $\mathrm{Rat}_d(\widehat{\mathbb{C}})$ of all rational maps, the bifurcation set has positive Lebesgue measure. Also, for non-trivial analytic families of rational maps, Shishikura [Sh2] and McMullen [McM3] proved that the Hausdorff dimension of the bifurcation set is equal to the dimension of the parameter space.

A much deeper understanding of the dynamics and bifurcation patterns has been obtained for the special family of quadratic polynomials $\{f_c(z) = z^2 + c \,|\, c \in \mathbb{C}\}$. On the one hand, for values of c outside the ball of radius 2 the iterates of the critical point 0 escape to infinity, and, as we shall see in section 3.3, the Julia set is a Cantor set and all the corresponding parameter values belong to the same topological conjugacy class. On the other hand, Douady and Hubbard proved in [DH2] that the set of parameter values for which the critical orbit is bounded, the so-called *Mandelbrot set* , is connected (see also [DH1] or [MNTU], pp. 21–2, for a proof) and also showed that its interior is a countable union of disjoint topological disks. Each of these disks, with the possible exception of an interior point that corresponds to a map having a periodic critical point, is a full conjugacy class. The bifurcation set of the quadratic family is the union of the boundary of the Mandelbrot set and the countable discrete set of maps with periodic critical points that lie in the interior of the Mandelbrot set. Significant progress in the understanding of the structure of this set, as well as of the Julia sets of quadratic polynomials, has been obtained in the works of Yoccoz, McMullen, Lyubich, Graczyk-Swiatek and others.

The mathematical tools that we will discuss in this book have been also very important in the study of the dynamics of circle and interval maps that are real analytic or even smooth; see [MS], [dFM2], [dFM1], and also [dFMP]. In this case the phase space reduces to a compact interval of the real line or to the circle, but the parameter space becomes an infinite-dimensional Banach space. Holomorphic methods still play an important role in the understanding of the small-scale structure of the orbits of smooth maps.

We conclude this introduction by mentioning some fundamental open problems. The most well-known, the so-called *Fatou conjecture*, states that for a structurally stable map each critical point is in the basin of an attracting periodic point. This is a very difficult problem, which is still open even for the quadratic family. For a long time it was conjectured that the Julia set of a rational map would either be the whole Riemann sphere or would have Lebesgue measure zero; in particular, the Julia set of every polynomial would have measure zero. This was recently disproved by X. Buff and A. Cheritat [BuC], who found Julia sets of positive Lebesgue measure in the quadratic family, a truly outstanding achievement. By M. Rees' theorem, the bifurcation set of the family of rational maps of degree d has positive Lebesgue measure. However, it is expected that the bifurcation set of the family of polynomials of degree d should have zero Lebesgue measure. For the quadratic family, an important conjecture formulated by Douady and Hubbard is that the Mandelbrot set is locally connected. They proved in [DH2] that this conjecture implies Fatou's conjecture for the quadratic family. Finally, a very important open question concerns the regularity of the conjugacy between two rational maps. It is conjectured that if two rational maps are topologically conjugate then a conjugacy exists between them that is quasiconformal. A solution to this conjecture in the special case of *real* quadratic polynomials, which implies the solution of Fatou's problem, was obtained in [L4] and in [GS1] and uses all the tools that we discuss in this book and more.

2
Preliminaries in complex analysis

Complex analysis is a vast and very beautiful subject, and the key to its beauty is the harmonious coexistence of analysis, algebra, geometry and topology in its most fundamental entity, the complex plane. We will assume that the reader is already familiar with the basic facts about analytic functions in one complex variable, such as Cauchy's theorem, the Cauchy–Riemann equations, power series expansions, residues and so on. Holomorphic functions in one complex variable enjoy a double life, as they can be viewed both as *analytic* objects (power series, integral representations) and as *geometric* objects (conformal mappings). The topics presented in this book exploit freely this dual character of holomorphic functions. Our purpose in this short chapter is to present some well-known or not so well-known analytic and geometric facts that will be necessary later. The reader is warned that what follows is only a brief collection of facts to be used, not a systematic exposition of the theory. For general background reading in complex analysis, see for instance [A2], [An] or [Rud].

2.1 Analytic facts

Let us start with some differential calculus of complex-valued functions defined on some domain in the complex plane (by a domain we mean as usual a non-empty, connected, open set). The two basic differential operators of complex calculus are

$$\frac{\partial}{\partial z} = \frac{1}{2}\left(\frac{\partial}{\partial x} - i\frac{\partial}{\partial y}\right) \; ; \; \frac{\partial}{\partial \bar{z}} = \frac{1}{2}\left(\frac{\partial}{\partial x} + i\frac{\partial}{\partial y}\right) ,$$

so that, if $f : \Omega \to \mathbb{C}$ ($\Omega \subseteq \mathbb{C}$ being a domain) is a C^1 function, then its total derivative is

$$df = \frac{\partial f}{\partial x} dx + \frac{\partial f}{\partial y} dy = \frac{\partial f}{\partial z} dz + \frac{\partial f}{\partial \overline{z}} d\overline{z} \,,$$

where $dz = dx + i\, dy$ and $d\overline{z} = dx - i\, dy$. We often simplify the notation even further, writing $\partial f = \partial f / \partial z$ and $\overline{\partial} f = \partial f / \partial \overline{z}$ respectively. Thus, a C^1 function is analytic, or *holomorphic*, if $\overline{\partial} f(z) = 0$ for all $z \in \Omega$. In this case the limit

$$f'(z) = \lim_{h \to 0} \frac{f(z + h) - f(z)}{h}$$

exists and equals $\partial f(z)$ for all $z \in \Omega$ (it is the complex derivative of f at z).

Several basic facts from standard calculus can be restated in complex notation. Thus, we can write Green's formula in the following way. If $u, v : \Omega \to \mathbb{C}$ are C^1 functions and $V \subseteq \Omega$ is a simply connected domain bounded by a piecewise C^1 Jordan curve (the boundary ∂V), then

$$\int_{\partial V} u\, dz + v\, d\overline{z} = \iint_V \left(\partial v - \overline{\partial} u \right) dz \wedge d\overline{z}; \tag{2.1}$$

here $dz \wedge d\overline{z} = (dx + i\, dy) \wedge (dx - i\, dy) = -2i\, dx \wedge dy$ is the complex area form.

With the help of Green's formula, it is not difficult to check that if $f : \Omega \to \mathbb{C}$ is a C^1 function and D is an open disk with $\overline{D} \subseteq \Omega$ then for all $z \in D$ we have

$$f(z) = \frac{1}{2\pi i} \int_{\partial D} \frac{f(\zeta)}{\zeta - z} d\zeta + \frac{1}{2\pi i} \iint_D \frac{\overline{\partial} f(\zeta)}{\zeta - z} d\zeta \wedge d\overline{\zeta}. \tag{2.2}$$

This is known as the *Cauchy–Green* or *Pompeiu* formula Note that if f is analytic then the second integral vanishes identically and we recover the usual Cauchy formula of basic complex analysis.

We can still make sense out of the preceding formulas even if the functions involved are not C^1. Indeed, we can think of ∂f or $\overline{\partial} f$ as *distributions*. The most useful situation occurs when the distributional derivatives $\partial f, \overline{\partial} f$ of a given $f : \Omega \to \mathbb{C}$ are represented by locally integrable functions $f_z, f_{\overline{z}} : \Omega \to \mathbb{C}$. In this case, for all test functions $\varphi \in C_0^\infty(\Omega)$ (complex-valued C^∞ functions with compact support) we have

$$\iint_\Omega f_z(\zeta) \varphi(\zeta)\, d\zeta \wedge d\overline{\zeta} = - \iint_\Omega f(\zeta) \partial \varphi(\zeta)\, d\zeta \wedge d\overline{\zeta} \,,$$

as well as

$$\iint_{\Omega} f_{\bar{z}}(\zeta)\varphi(\zeta)\,d\zeta \wedge d\bar{\zeta} \;=\; -\iint_{\Omega} f(\zeta)\overline{\partial}\varphi(\zeta)\,d\zeta \wedge d\bar{\zeta}.$$

When working with the distributional derivatives of a given f, as above, we often need to approximate f in a suitable sense by a sequence of smooth functions usually referred to as a *smoothing sequence*. Such an approximation is obtained by performing the convolution of f with an *approximate identity*, a sequence of C^{∞} functions $\phi_n : \mathbb{C} \to \mathbb{C}$ with compact support having the following properties:

(1) $\displaystyle\iint_{\mathbb{C}} |\phi_n(z)|\,dxdy = 1$

(2) $\operatorname{supp}\phi_n \subset D(0, 1/n)$.

The standard example of an approximate identity is constructed as follows. Let $\varphi : \mathbb{C} \to \mathbb{R}$ be the C^{∞} function which is given by

$$\varphi(z) \;=\; \exp\left(-\frac{1}{1 - |z|^2}\right)$$

for all $z \in \mathbb{D}$ and which vanishes identically outside $\mathbb{D}$. Let $\phi : \mathbb{C} \to \mathbb{R}$ be defined by $\phi(z) = \varphi(z)/\iint_{\mathbb{C}} |\varphi(z)|\,dxdy$, and then take $\phi_n(z) = n^2\phi(nz)$ for each $n \geq 1$. Now we have the following important fact.

Lemma 2.1.1 *Let $f : \Omega \to \mathbb{C}$ be a continuous function whose distributional derivatives $f_z, f_{\bar{z}}$ are such that $|f_z|^p$ and $|f_{\bar{z}}|^p$ are locally integrable on Ω, for some fixed $p \geq 1$. Then for each compact set $K \subset \Omega$ there exists a sequence $f_n \in C_0^{\infty}(\Omega)$ such that f_n converges uniformly to f on K as $n \to \infty$ and such that*

$$\lim_{n\to\infty} \iint_{K} |\partial f_n(z) - f_z(z)|^p\,dxdy \;=\; 0$$

as well as

$$\lim_{n\to\infty} \iint_{K} |\overline{\partial} f_n(z) - f_{\bar{z}}(z)|^p\,dxdy \;=\; 0\,.$$

Such a sequence is called an L^p smoothing sequence for f in K.

Proof Consider a fixed $\lambda_K \in C_0^{\infty}(\Omega)$ that is constant and equal to 1 on some neighborhood of K, and form the function $f_K = \lambda_K f$, which has compact support in K; extend f_K outside Ω, setting it equal to zero.

Note that the distributional partial derivatives of f_K are locally in L^p; in fact

$$\partial f_K = \partial \lambda_K f + \lambda_K \partial f , \quad \overline{\partial} f_K = \overline{\partial} \lambda_K f + \lambda_K \overline{\partial} f .$$

Let $\{\phi_n\}_{n \geq 1}$ be an approximate identity (say the one we constructed before lemma 2.1.1), and let $f_n \in C_0^\infty(\Omega)$ be given by

$$f_n(z) = \phi_n * f_K(z) = -\frac{1}{2i} \iint_{\mathbb{C}} \phi_n(z - \zeta) f_K(\zeta) \, d\zeta \wedge d\overline{\zeta} .$$

Then we have also

$$\partial f_n = \phi_n * \partial f_K , \qquad \overline{\partial} f_n = \phi_n * \overline{\partial} f_K .$$

It now follows from standard properties of convolutions that $f_n \to f_K$ uniformly in K and that $\partial f_n \to \partial f_K$ and $\overline{\partial} f_n \to \overline{\partial} f_K$ in $L^p(K)$. This is the desired result, because for all $z \in K$ we have $f_K(z) = f(z)$, $\partial f_K(z) = f_z(z)$ and $\overline{\partial} f_K(z) = f_{\overline{z}}(z)$. $\qquad \square$

This lemma yields two key results. The first is a fundamental lemma due to H. Weyl, which is a special case of a much more general regularity theorem for elliptic operators.

Proposition 2.1.2 (Weyl's lemma) *If $f : \Omega \to \mathbb{C}$ is a continuous function such that $\overline{\partial} f = 0$ in Ω in the sense of distributions then f is holomorphic in Ω.*

Proof Take any disk D whose closure is contained in Ω. Let $f_n : \Omega \to \mathbb{C}$ be an L^1 smoothing sequence for f in $\overline{D}$. Then f_n converges uniformly to f in $\overline{D}$. From the fact that $\overline{\partial} f = 0$ in the distributional sense, it follows that $\overline{\partial} f_n(z) = 0$ for all $z \in D$. Since f_n is C^1, it follows that f_n is holomorphic for each n. Therefore f, being the uniform limit of holomorphic functions, is holomorphic in D. Since $D \subset \Omega$ is arbitrary, f is holomorphic in Ω. $\qquad \square$

The second result that we prove with the help of lemma 2.1.1 is the following more general version of (2.2).

Proposition 2.1.3 (Pompeiu's formula) *Let $f : \Omega \to \mathbb{C}$ be a continuous function whose distributional derivatives $\partial f, \overline{\partial} f$ are represented by functions $f_z, f_{\overline{z}}$ locally in L^p for some fixed p with $2 < p < \infty$. Then for each open disk D compactly contained in Ω and each $z \in D$ we have*

$$f(z) = \frac{1}{2\pi i} \int_{\partial D} \frac{f(\zeta)}{\zeta - z} \, d\zeta + \frac{1}{2\pi i} \iint_D \frac{f_{\overline{z}}(\zeta)}{\zeta - z} \, d\zeta \wedge d\overline{\zeta} . \qquad (2.3)$$

Proof Note that the last integral is absolutely convergent because $f_{\bar{z}} \in L^p(D)$ and $1/(\zeta - z) \in L^q(D)$, where $q < 2$ is the conjugate exponent of p (that is, $p^{-1} + q^{-1} = 1$). Let $f_n \in C_0^\infty(\Omega)$ be an L^p smoothing sequence for f in D. By (2.2), for each n we have

$$f_n(z) = \frac{1}{2\pi i} \int_{\partial D} \frac{f_n(\zeta)}{\zeta - z} \, d\zeta + \frac{1}{2\pi i} \iint_D \frac{\overline{\partial} f_n(\zeta)}{\zeta - z} \, d\zeta \wedge d\bar{\zeta} .$$

Since $f_n \to f$ uniformly in $\overline{D}$, whereas $\overline{\partial} f_n \to f_{\bar{z}}$ in $L^p(\overline{D})$, we deduce (say by the dominated convergence theorem) that (2.3) holds. □

2.2 Geometric inequalities

The theory of conformal mappings is extremely rich in inequalities having a geometric content.

2.2.1 The classical Schwarz lemma

The most fundamental inequality in complex function theory is the classical *Schwarz lemma*, which we now recall. It states that every holomorphic self-map of the unit disk that fixes the origin is either a contraction near the origin or else it is a rotation. The precise statement is the following.

Lemma 2.2.1 (Schwarz) *Let* $f : \mathbb{D} \to \mathbb{D}$ *be a holomorphic map such that* $f(0) = 0$, *and let* $\lambda = f'(0)$. *Then either* $|\lambda| < 1$, *in which case* $|f(z)| < |z|$ *for all* z, *or else* $|\lambda| = 1$, *in which case* $f(z) = \lambda z$ *for all* z.

Proof Let $\varphi : \mathbb{D} \to \mathbb{D}$ be given by

$$\varphi(z) = \begin{cases} \dfrac{f(z)}{z} & \text{if } z \in \mathbb{D} \setminus \{0\}, \\[2ex] f'(0) & \text{if } z = 0 . \end{cases}$$

By Riemann's removable singularity theorem, φ is holomorphic. Hence, for all $z \in \mathbb{D}$ and all r such that $|z| \le r < 1$ we have, by the maximum principle,

$$|\varphi(z)| \le \sup_{|\zeta|=r} \left| \frac{f(\zeta)}{\zeta} \right| \le \frac{1}{r} .$$

Letting $r \to 1$, we deduce that $|\varphi(z)| \le 1$, i.e. $|f(z)| \le |z|$, for all $z \in \mathbb{D}$. If equality holds for some z then, again by the maximum principle, φ

must be constant, this constant can only be $\lambda = \varphi(0)$, and of course in this case $f(z) = \lambda z$ for all z. □

The Schwarz lemma has many other formulations, some of which will be given in the next chapter (see lemma 3.1.4 and theorem 3.3.1). Here we present an invariant form of the Schwarz lemma due to G. Pick. First note that fractional linear transformations (FLTs) of the form

$$T(z) = \frac{az + b}{\bar{b}z + \bar{a}}, \qquad (2.4)$$

with $|a|^2 - |b|^2 = 1$, map the unit disk onto itself bi-holomorphically, their inverses being of the same type. They are complex automorphisms of the unit disk and form a Lie group isomorphic to the matrix group $PSU(1,1) = SU(1,1)/\{I, -I\}$, where

$$SU(1,1) = \left\{ \begin{pmatrix} a & b \\ \bar{b} & \bar{a} \end{pmatrix} : a, b \in \mathbb{C}, \ |a|^2 - |b|^2 = 1 \right\}.$$

Pick's formulation of the Schwarz lemma says that these maps comprise *all* the automorphisms of $\mathbb{D}$.

Lemma 2.2.2 (Schwarz–Pick) *Let $f : \mathbb{D} \to \mathbb{D}$ be holomorphic. Then for all $z, w \in \mathbb{D}$ we have*

$$\left| \frac{f(z) - f(w)}{1 - f(z)\overline{f(w)}} \right| \leq \left| \frac{z - w}{1 - z\overline{w}} \right|, \qquad (2.5)$$

and equality holds if and only if f is an FLT of the form (2.4).

Proof Let $T : \mathbb{D} \to \mathbb{D}$ be the FLT given by

$$T(\zeta) = \frac{\zeta - w}{1 - \overline{w}\zeta},$$

and let $S : \mathbb{D} \to \mathbb{D}$ be the FLT given by

$$S(\zeta) = \frac{\zeta - f(w)}{1 - \overline{f(w)}\zeta}.$$

Then $g = S \circ f \circ T^{-1} : \mathbb{D} \to \mathbb{D}$ satisfies $g(0) = 0$. Therefore by Schwarz's lemma we have $|g(\zeta)| \leq |\zeta|$ for all ζ, and in particular $|g(T(z))| \leq |T(z)|$. But this is exactly what (2.5) says. □

The Schwarz–Pick lemma shows in particular that the group $\mathrm{Aut}(\mathbb{D})$ of complex automorphisms of the unit disk is isomorphic to $PSU(1,1)$.

Note that if we divide both sides of (2.5) by $|z - w|$ and let $w \to z$, we get the inequality

$$\frac{|f'(z)|}{1 - |f(z)|^2} \leq \frac{1}{1 - |z|^2} . \tag{2.6}$$

This suggests that we consider the conformal metric ds^2 in the unit disk given by

$$ds = \frac{2|dz|}{1 - |z|^2} .$$

This is a Riemannian metric in $\mathbb{D}$ with constant Gaussian curvature equal to -1. It is called the *hyperbolic metric* of the unit disk. The inequality (2.6) tells us that every holomorphic self-map of the unit disk is either an isometry in this metric or else it strictly contracts it. The (orientation-preserving) isometries of the hyperbolic metric of the disk are precisely the FLT's that form Aut($\mathbb{D}$). For a more geometric approach to hyperbolic geometry, see the next chapter.

2.2.2 Koebe's one-quarter theorem

One knows from the open mapping theorem (see [A2]) that the image, under an analytic function f, of a disk around a point z in a domain always contains a disk around the image point $f(z)$. What can we say about the relative (maximal) sizes of such disks? Without further assumptions on f the answer is, not much (see exercise 2.4). However, if the function f is *univalent* (near z) then there is an estimate on the relative sizes of such disks. By univalent we mean *injective*. The estimate is given by Koebe's *one-quarter theorem*, which will be proved below.

Before we state Koebe's theorem, let us introduce another result which will be used in the proof and which is of independent interest (see Chapter 7). This result requires for the following basic fact, left as an exercise to the reader. If γ is a (positively oriented) simple closed curve of class C^1 in the complex plane, bounding a topological disk D, then

$$\text{Area}\, D = \frac{1}{2i} \int_\gamma \overline{z}\, dz .$$

The theorem below was first proved by L. Bieberbach [Bi], and it is known as the *area theorem* or *Bieberbach's flächensatz*.

Theorem 2.2.3 (Area theorem) *Let φ be an injective analytic function from the punctured unit disk $0 < |z| < 1$ into the complex plane, and suppose that*

$$\varphi(z) = \frac{1}{z} + b_0 + b_1 z + b_2 z^2 + \cdots + b_n z^n + \cdots . \qquad (2.7)$$

Then $\sum_{n=1}^{\infty} n|b_n|^2 \leq 1$.

Proof Take $0 < r < 1$ and let D_r^* be a disk of radius r around the origin that is punctured at the origin. Let $\Omega_r = \mathbb{C} \setminus \varphi(D_r)$. Then, since φ is injective, Ω_r is a topological disk bounded by the Jordan curve $\gamma_r = \varphi(\partial D_r)$. Hence

$$\text{Area}\,\Omega_r = \frac{1}{2i} \int_{\gamma_r} \bar{w}\, dw = -\frac{1}{2i} \int_{|z|=r} \overline{\varphi(z)} \varphi'(z)\, dz ,$$

where the minus sign accounts for the counterclockwise orientation of γ_r. Using the Laurent expansion (2.7) in this last expression, we get

$$\text{Area}\,\Omega_r = -\frac{1}{2i} \int_{|z|=r} \left(\frac{1}{\bar{z}} + \sum_{n=0}^{\infty} \overline{b_n}\, \bar{z}^n \right) \left(-\frac{1}{z^2} + \sum_{n=1}^{\infty} n b_n z^{n-1} \right) dz .$$
$$(2.8)$$

Taking into account that

$$\int_{|z|=r} \bar{z}^m z^{n-1}\, dz = \begin{cases} 2\pi r^{2n} & \text{if } m = n, \\ \\ 0 & \text{if } m \neq n , \end{cases}$$

we deduce from (2.8) that

$$\text{Area}\,\Omega_r = \pi \left(\frac{1}{r^2} - \sum_{n=1}^{\infty} n|b_n|^2 r^{2n} \right) .$$

But the area on the left-hand side is a non-negative number; hence for all $0 < r < 1$ we must have

$$\sum_{n=1}^{\infty} n|b_n|^2 r^{2n} \leq \frac{1}{r^2} .$$

Letting $r \to 1$ in this last inequality, we deduce that $\sum_{n=1}^{\infty} n|b_n|^2 \leq 1$ as claimed.

$\square$

The area theorem has the following consequence.

Lemma 2.2.4 *If $f(z) = z + a_2 z^2 + \cdots$ is univalent in the unit disk then:*

    (i) *there exists $g : \mathbb{D} \to \mathbb{C}$, also univalent, such that $[g(z)]^2 = f(z^2)$ for all $z \in \mathbb{D}$;*

    (ii) *we have $|a_2| \le 2$.*

Proof First define $\varphi(z) = f(z)/z$ for $z \ne 0$, and $\varphi(0) = 1$. Then φ is holomorphic (check!) and vanishes nowhere in $\mathbb{D}$ because f is injective. Therefore there exists $h : \mathbb{D} \to \mathbb{C}$, which is also holomorphic, such that $(h(z))^2 = \varphi(z)$ (see exercise 2.2). Let g be given by $g(z) = zh(z^2)$; then

$$\left(g(z)\right)^2 = z^2 \left(h(z^2)\right)^2 = z^2 \varphi(z^2) = f(z^2) \, .$$

Hence to prove (i) it suffices to show that g is univalent. Suppose that $z, w \in \mathbb{D}$ are such that $g(z) = g(w)$. Then $f(z^2) = f(w^2)$, so $z^2 = w^2$ because f is univalent. Thus, either $z = w$ or $z = -w$. But since g is clearly an odd function, if $z = -w$ then $g(z) = -g(w)$ and therefore $g(z) = 0 = g(w)$. But this is only possible if $z = 0 = w$ (because h is nowhere zero). This shows that in any case we have $z = w$, so g is univalent. This proves (i).

To prove (ii), let $G : \mathbb{D}^* \to \mathbb{C}$ be given by $G(z) = 1/g(z)$. Note that G is injective, because g is injective. Since

$$f(z^2) = z^2(1 + a_2 z^2 + \cdots) \, ,$$

we have

$$g(z) = z(1 + \tfrac{1}{2}a_2 z^2 + \cdots) \, .$$

This shows that the Laurent expansion of G is

$$G(z) = \frac{1}{z} - \tfrac{1}{2}a_2 z + \cdots \, .$$

Applying the area theorem, we get

$$\left|-\tfrac{1}{2}a_2\right| \le 1 \, .$$

Therefore $|a_2| \le 2$, and this finishes the proof.

$\square$

Now it is straightforward to deduce Koebe's one-quarter theorem from the above lemma.

Theorem 2.2.5 (Koebe's one-quarter theorem) *Let f be a univalent map from the unit disk into the complex plane, and suppose that $f(0) = 0$ and $|f'(0)| = 1$. Then the image of f contains a disk of radius $1/4$ about the origin.*

Proof Let us write $f(z) = z + a_2 z^2 + \cdots$; from the previous lemma, we know that $|a_2| \leq 2$. Suppose $w \in \mathbb{C}$ is not in the image of f, and let

$$F(z) = \frac{w f(z)}{w - f(z)} .$$

Then F is holomorphic and univalent in the unit disk and the Taylor series of F is

$$F(z) = z + \left(a_2 + \frac{1}{w}\right) z^2 + \cdots$$

Hence F also satisfies the hypotheses of our previous lemma, so that

$$\left| a_2 + \frac{1}{w} \right| \leq 2 .$$

Therefore $|w^{-1}| \leq 2 + |a_2| \leq 4$, so that $|w| \geq 4$. But this means that the disk of radius $1/4$ about the origin is entirely contained in the image of f. $\qquad\square$

2.3 Normal families

In complex function theory, one often deals with families of holomorphic functions. To extract limits from sequences of functions in these families, one needs some form of (sequential) compactness. This leads naturally to the notion of a *normal family*.

Definition 2.3.1 *A family $\mathcal{F}$ of holomorphic functions defined over a fixed domain $V \subseteq \mathbb{C}$ is said to be normal if every sequence of members of $\mathcal{F}$ has a subsequence that converges uniformly on compact subsets of V.*

Note that we do not require the limits of such subsequences to belong to $\mathcal{F}$. The following basic result on normal families was established by P. Montel (who actually proved a much stronger result, see Chapter 3).

Theorem 2.3.2 (Montel) *If a family $\mathcal{F}$ of holomorphic functions over a fixed domain V is uniformly bounded on compacta, i.e. if for each* compact subset $K \subset V$ there exists $M_K > 0$ such that $\sup_{z \in K} |f(z)| \leq M_K$ for all $f \in \mathcal{F}$, then $\mathcal{F}$ is normal.

Proof This theorem follows from the Arzelá–Ascoli theorem if one can prove that the family is also equicontinuous on compacta. Given a compact subset $K \subset V$, let $\delta_K = \text{dist}(K, \partial V)/3$, and consider

$$K^* = \{z \,:\, \text{dist}(z, K) \leq 2\delta_K\} \subset V \,.$$

Then K^* is also compact. If z_1 and z_2 are any two points of K such that $|z_1 - z_2| < \delta_K$ then the closed disk D of center z_1 and radius $2\delta_K$ contains Z_2 and is contained in K^*. Applying Cauchy's integral formula, for every $f \in \mathcal{F}$ we have

$$f(z_1) - f(z_2) = \frac{z_1 - z_2}{2\pi i} \int_{\partial D} \frac{f(\zeta)}{(\zeta - z_1)(\zeta - z_2)} \, d\zeta \,. \qquad (2.9)$$

But since $|\zeta - z_1| = 2\delta_K$ and $|\zeta - z_2| > \delta_K$ for all $\zeta \in \partial D$, we deduce from (2.9) that

$$|f(z_1) - f(z_2)| \leq \frac{M_{K^*}}{\delta_K} |z_1 - z_2| \,.$$

This proves that $\mathcal{F}$ is equicontinuous on K.

$\qquad\qquad\qquad\qquad\qquad\qquad\qquad\qquad\qquad\qquad\qquad\qquad\qquad\qquad\square$

One of the first and most important applications of Montel's theorem was to the proof of the Riemann mapping theorem, according to which every proper simply connected subdomain of the complex plane is conformally equivalent to the unit disk (a short proof can be found in [Rud, pp. 302–4]). Again, see Chapter 3, where the more general uniformization theorem for domains in the Riemann sphere will be proved.

Exercises

2.1   Find all complex automorphisms of the upper half-plane $\mathbb{H} = \{z \in \mathbb{C} : \text{Im}\, z > 0\}$.

2.2   Let $V \subset \mathbb{C}$ be simply connected, and let $g : V \to \mathbb{C}$ be analytic. Show that, if $g(z) \neq 0$ for all $z \in V$, there exists an analytic function $h : V \to \mathbb{C}$ such that $g(z) = (h(z))^2$.

2.3   Show that the function

$$f(z) = \frac{z}{(1 - z)^2}$$

is univalent in the unit disk, and find its image $f(\mathbb{D})$. Use this f to show that the constant $1/4$ in Koebe's theorem is sharp. What is the Taylor expansion of f about zero?

2.4   Koebe's theorem can be weakened to the following qualitative statement: if f is a holomorphic normalized univalent map of the unit disk then its image contains a disk of *definite* radius centered at the origin. Show that this statement becomes false if the hypothesis that f is univalent is removed. (*Hint* Consider $f_n(z) = (e^{nz} - 1)/n$.)

2.5   Let f be analytic in the unit disk but not necessarily univalent. Suppose that $f(0) = 0$, $|f'(0)| = 1$ and $|f(z)| \le M$ for all $z \in \mathbb{D}$. Show that $f(\mathbb{D})$ contains the disk of radius $1/(4M)$ about zero.

2.6   This exercise outlines a proof of a weak version of what is known as *Koebe's distortion theorem* (for the full theorem, see [CG, p. 3]).

(a) Let $F : \mathbb{D} \to \mathbb{C}$ be a normalized univalent function. Show that

$$\left| \frac{F''(0)}{F'(0)} \right| \le 4 \ .$$

(b) Let $f : \mathbb{D} \to \mathbb{C}$ be univalent, let $\zeta \in \mathbb{D}$ and consider $F(z) = \lambda_\zeta (f \circ T_\zeta(z) - f(\zeta))$, where

$$T_\zeta(z) \ = \ \frac{z + \zeta}{1 + \bar\zeta z}$$

and $\lambda_\zeta = f'(\zeta)(1 - |\zeta|^2)$. Show that F is a normalized univalent function, and compute $F''(0)/F'(0)$.

(c) Deduce from (a) and (b) that

$$\left| \frac{f''(\zeta)}{f'(\zeta)} \right| \ \le \ \frac{4}{\text{dist}(\zeta, \partial \mathbb{D})} \ ,$$

where dist denotes the Euclidean distance.

2.7   Prove *Vitali's theorem*, as follows. Let $f_n : V \to \mathbb{C}$ be a uniformly bounded sequence of analytic functions on a connected open set $V \subseteq \mathbb{C}$, and suppose that this sequence converges pointwise on a subset $E \subset V$ having at least one accumulation point in V. Then (f_n) converges uniformly on compact subsets to some analytic function $f : V \to \mathbb{C}$.

3

Uniformization and conformal distortion

The most fundamental result of Riemann surface theory is the *uniformization theorem*, which states that the universal covering space of every Riemann surface is isomorphic to either the complex plane, the Riemann sphere or the unit disk. In particular, every domain in the Riemann sphere whose complement has at least three points is covered by the unit disk. The natural hyperbolic metric on the disk can be pushed down to any such domain via the covering projection. Thus every domain in the Riemann sphere whose complement has at least three points has a natural hyperbolic structure, and this has many useful consequences. For instance, there is a version of Schwarz's lemma for holomorphic maps between such domains. In this chapter, our discussion of uniformization is restricted to these domains in the Riemann sphere, because this is quite sufficient for the dynamical applications we have in mind. For the general uniformization theorem, the reader may consult [FK] (and also the appendix). We also discuss conformal distortion tools arising from Koebe's distortion theorem. The applications we present in this chapter include the basics of the Fatou–Julia theory regarding the iteration of rational maps and the thermodynamic formalism for Cantor repellers.

3.1 The Möbius group

Recall that the Riemann sphere $\widehat{\mathbb{C}}$ is the union of the complex plane $\mathbb{C}$ and the point ∞. The neighborhoods of a finite point are the usual ones, whereas the neighborhoods of ∞ are the complements of the compact subsets of $\mathbb{C}$. If $U \subset \widehat{\mathbb{C}}$ is an open neighborhood of ∞ then we say that a function $f: U \to \mathbb{C}$ is holomorphic at ∞ if the function $z \mapsto f(1/z)$, $z \neq$

$0, 0 \mapsto f(\infty)$ is holomorphic at 0. Similarly, we can define holomorphic functions whose range is the Riemann sphere.

Three special types of maps of the Riemann sphere deserve to be listed:

(i) the conformal involution $I(z) = z^{-1}$ about the unit circle;

(ii) the homotheties $M_a(z) = az$, where $a \neq 0$ is a complex number;

(iii) the translations $T_b(z) = z + b$, where b is any complex number.

These are holomorphic diffeomorphisms of the Riemann sphere. Hence, any map which is a finite composition of maps in this list is also a holomorphic diffeomorphism of the Riemann sphere. These maps form a group, under the operation of composition, which is denoted by $\text{Möb}(\widehat{\mathbb{C}})$ and is called the *Möbius group*.

Let $\mathcal{C}$ be the set of all Euclidean circles of the Riemann sphere. An element of $\mathcal{C}$ is either a straight line in the plane, if it contains the point ∞, or an Euclidean circle in the complex plane. Three distinct points z_1, z_2, z_3 of the Riemann sphere determine a unique element of $\mathcal{C}$. Since each generator of the Möbius group preserves the family $\mathcal{C}$, it follows that same holds for all elements of the Möbius group.

Given any three distinct points z_1, z_2, z_3 of the Riemann sphere, there exists a unique Möbius transformation ϕ that maps z_1 to $0, z_2$ to 1, z_3 to ∞. In fact, the map $\psi_1 = I \circ T_{-z_3}$ maps z_1 to w_1, z_2 to w_2, with $w_1 \neq w_2$, and z_3 to ∞. The Möbius transformation

$$\psi_2 = M_{1/(w_2-w_1)} \circ T_{-w_1}$$

fixes ∞ and maps w_1 to 0 and w_2 to 1. The composition $\psi_2 \circ \psi_1$ is the desired ϕ. Uniqueness follows from the fact that a holomorphic diffeomorphism ψ that fixes $0, 1, \infty$ must be the identity, since the map $z \mapsto \psi(z)/z$ is holomorphic in the whole sphere, does not vanish and is therefore constant. Thus a Möbius transformation is characterized by the image of any given three distinct points. As a consequence, any holomorphic diffeomorphism of the Riemann sphere must be a Möbius transformation and hence may be written as a composition of at most four of the above generators of the group. It follows also that the Möbius group acts transitively on $\mathcal{C}$: given $C_1, C_2 \in \mathcal{C}$, there exists a Möbius transformation ϕ such that $\phi(C_1) = C_2$. Sometimes it is useful to consider a larger group, that of all conformal diffeomorphisms of the Riemann sphere. To generate this group it is enough to add to the previous set of generators the *geometric* inversion with respect to the unit circle, namely the map

$$\overline{I}(z) \; = \; \frac{z}{|z|^2} \; .$$

This orientation-reversing diffeomorphism fixes every point in the boundary of the unit disk $\mathbb{D}$. Note that $\overline{I}$ is simply the involution I post-composed with complex conjugation. If $C \in \mathcal{C}$ and ϕ is a Möbius transformation mapping $\partial\mathbb{D}$ onto C then $\overline{I}_C = \phi \circ \overline{I} \circ \phi^{-1}$ is the (geometric) inversion with respect to the circle C. It is easy to verify that each generator of the Möbius group is the composition of two inversions. Therefore any Möbius transformation is a composition of inversions.

The group structure of $\mathrm{M\ddot{o}b}(\widehat{\mathbb{C}})$ may be conveniently described by the group $SL(2, \mathbb{C})$ of complex 2×2 matrices having determinant equal to 1. Indeed, it is easy to verify that the map

$$SL(2, \mathbb{C}) \ni \begin{pmatrix} a & b \\ c & d \end{pmatrix} \longmapsto \left\{ z \mapsto \frac{az + b}{cz + d} \right\} \in \mathrm{M\ddot{o}b}(\widehat{\mathbb{C}})$$

is a group homomorphism onto $\mathrm{M\ddot{o}b}(\widehat{\mathbb{C}})$, i.e. the composition of Möbius transformations corresponds to matrix multiplication. Also, the kernel of the homomorphism is just the subgroup $\{-I, I\}$, where I is the identity matrix. Hence $\mathrm{M\ddot{o}b}(\widehat{\mathbb{C}}) \cong SL(2, \mathbb{C})/\{-I, I\} = PSL(2, \mathbb{C})$.

The dynamics of a Möbius transformation is very simple to describe. First, there are only two possibilities concerning fixed points: a Möbius transformation has at least one and at most two fixed points. Suppose the Möbius transformation ϕ has two fixed points. Take another Möbius transformation ψ that maps the fixed points of ϕ to 0 and ∞. Then the Möbius transformation $\psi^{-1} \circ \phi \circ \psi^{-1}$ has 0 and ∞ as fixed points. Therefore it is of the form $z \mapsto az$, where a is a non-zero complex number. If $|a| \neq 1$ then one fixed point is attracting and the other is repelling, and any other orbit converges to the attracting (repelling) fixed point under positive (negative) iteration. These maps are called *loxodromic*, or *hyperbolic* if a is real. If $|a| = 1$, the Möbius transformation is a rotation, and either all other orbits will be periodic if a is a root of unity or each will be dense in a circle if a is not a root of unity. These maps are called *elliptic*. If there is only one fixed point, we can choose ψ to map this fixed point to ∞ in such a way that $\psi \circ \phi \circ \psi^{-1}$ is the translation $z \mapsto z + 1$. In this case all other orbits will be asymptotic to the fixed point both under positive and under negative iterations. Such maps are called *parabolic*.

Let $\mathbb{D} = \{z \in \mathbb{C} : |z| < 1\}$ be the unit disk and $\mathbb{H} = \{z \in \mathbb{C} : \mathrm{Im}\, z > 0\}$ be the upper half-plane. We denote by $\mathrm{M\ddot{o}b}(\mathbb{D})$ ($\mathrm{M\ddot{o}b}(\mathbb{H})$) the subgroup of all Möbius transformations that preserve $\mathbb{D}$ ($\mathbb{H}$). The Möbius transformation ϕ that maps the points $-1, 1, i$ in the boundary of $\mathbb{D}$ respectively to the points $-1, 1, \infty$ in the boundary of $\mathbb{H}$ is a holomorphic diffeomorphism between $\mathbb{D}$ and $\mathbb{H}$. Hence $\phi^{-1}\text{Möb}(\mathbb{H})\phi = \text{Möb}(\mathbb{D})$ and the two subgroups are conjugate. For each $z \in \mathbb{D}$ ($z \in \mathbb{H}$) let us denote by $\mathcal{D}_z$ ($\mathcal{H}_z$) the set of all circles in $\mathcal{C}$ that are orthogonal to the boundary of $\mathbb{D}$ ($\mathbb{H}$) and that contain the point z. In particular, $\mathcal{D}_0$ is the set of all straight lines through the origin. It is clear that the above Möbius transformation ϕ establishes a bijection between $\mathcal{D}_z$ and $\mathcal{H}_{\phi(z)}$.

Proposition 3.1.1 *The following statements hold true.*

(i) *Given circles $C_i \in \mathcal{C}_{z_i}$, $z_i \in \mathbb{D}$, $i = 1, 2$, there exist exactly two elements of* $\text{Möb}(\mathbb{D})$ *mapping C_1 onto C_2 and z_1 to z_2. If $z_1 = z_2$ and $C_1 = C_2$ then one of these Möbius transformations is the identity and the other permutes the points of intersection of C_1 with the boundary of $\mathbb{D}$.*

(ii) *If $\psi \colon \mathbb{D} \to \mathbb{D}$ is an automorphism of $\mathbb{D}$ then ψ belongs to $\text{Möb}(\mathbb{D})$.*

(iii) *The inversion with respect to any circle orthogonal to the boundary of $\mathbb{D}$ maps $\mathbb{D}$ onto $\mathbb{D}$. Any element of $\text{Möb}(\mathbb{D})$ is a composition of such inversions.*

Proof Let $\{z_i^+, z_i^-\}$ be the intersection of C_i with the boundary of $\mathbb{D}$. The Möbius transformation that maps the points z_1^-, z_1, z_1^+ respectively to z_2^-, z_2, z_2^+ maps $\mathbb{D}$ onto $\mathbb{D}$ and C_1 onto C_2. The Möbius transformation that maps z_1^-, z_1, z_1^+ respectively to z_2^+, z_2, z_2^- also has the same properties, and the first part of the proposition follows. On the one hand, the rotations are elements of $\text{Möb}(\mathbb{D})$ that fix zero. On the other hand, by Schwarz's lemma, any automorphism of $\mathbb{D}$ that fixes the origin must be a rotation. Now, if $f \colon \mathbb{D} \to \mathbb{D}$ is an automorphism then, by the first part of the proposition, there exists $\phi \in \text{Möb}(\mathbb{D})$ such that $\phi(f(0)) = 0$. Hence $\phi \circ f$ is a rotation and the second part of the proposition is proved. To prove the last part, let us consider the group $\text{Inv}(\mathbb{D})$ generated by all inversions across the circles that are orthogonal to $\partial\mathbb{D}$. A rotation of angle θ about the origin is easily seen to be the composition of complex conjugation with an inversion about the straight line through the origin with angle $\theta/2$. Thus, every rotation about the origin is an element of $\text{Inv}(\mathbb{D})$. Next, note that if p is any point of $\mathbb{D}$, there is an inversion that maps the origin to p. Indeed, consider the straight line L that passes through p and is orthogonal to the straight line that joins p to the origin; the line L intersects $\partial\mathbb{D}$ in two points, say p' and p''. If

C is the circle that passes through p' and p'' and is orthogonal to the boundary of the unit disk, then one easily sees that the origin and the point p are mapped to each other by an inversion through C. Hence, if ϕ is an arbitrary element of Möb($\mathbb{D}$) and if $p = \phi(0)$, we can compose ϕ with an inversion that maps p to 0 and obtain a new Möbius transformation which fixes the origin and which must therefore be a rotation. This shows that ϕ is the composition of an inversion with a rotation about the origin and, since every such rotation is already in Inv($\mathbb{D}$), we see that ϕ also belongs to the group Inv($\mathbb{D}$). Since ϕ is arbitrary, this shows that Möb($\mathbb{D}$) = Inv($\mathbb{D}$) as asserted. □

Let $U \subset \mathbb{C}$ be an open subset. A *Riemannian metric* on U is a map that associates with each $z \in U$ an inner product $\langle \cdot, \cdot \rangle_z$ of $\mathbb{C}$ such that if $X, Y : U \to \mathbb{C}$ are C^∞ vector fields then $z \mapsto \langle X(z), Y(z) \rangle_z$ is a C^∞ function. The *length* of a piecewise differentiable curve $\gamma : [0, 1] \to U$ is defined by

$$\ell(\gamma) = \int_0^1 |\gamma'(t)|_{\gamma(t)} \, dt \ ,$$

where $|v|_z = \sqrt{\langle v, v \rangle_z}$ is the norm of the vector v at the point z. Given a Riemannian metric on an open set U, the area of the unit square with vertices $0, 1, 1 + i, i$ with respect to the inner product $\langle \cdot, \cdot \rangle_z$ at the point $z = x + iy$ is equal to

$$\sigma(x, y) = \sqrt{\langle 1, 1 \rangle_z \langle i, i \rangle_z - (\langle 1, i \rangle_z)^2}$$

and therefore the area of a domain $D \subset U$ is

$$A(D) = \iint_D \sigma(x, y) \, dx dy \ .$$

We say that the Riemannian metric is *conformal* if, for each z, $\langle v, w \rangle_z = (\rho(z))^2 \langle v, w \rangle$ where $\langle \cdot, \cdot \rangle$ is the Euclidean inner product of $\mathbb{C}$ and ρ is a C^∞ positive function. In this case, the length of a curve is given by

$$\ell(\gamma) = \int_0^1 \rho(\gamma(t)) |\gamma'(t)| \, dt \ ,$$

and the area of a region $D \subset U$ is defined as

$$A(D) = \iint_D \rho^2 \, dx dy \ .$$

If a curve has minimal length among all piecewise differentiable curves with the same endpoints then it is a *geodesic* of the metric. More generally, a geodesic is a curve that satisfies this property in some neighborhood of each of its points, i.e. for each t_0 there exists $\epsilon > 0$ such that $\gamma|[t_0, t_0 + \epsilon]$ has minimal length among all piecewise differentiable curves with the same endpoints. One can prove that geodesics are the solutions of a certain second-order ordinary differential equation. From the existence and uniqueness theorem for differential equations it follows that given $z \in U$ and a vector $v \in \mathbb{C}$ there exists a unique geodesic passing through z and whose tangent vector at this point is v. An *isometry* is a diffeomorphism $f\colon U \to U$ such that $\langle v, w \rangle_z = \langle Df(z)v, Df(z)w \rangle_{f(z)}$ for all $z \in U$ and all $v, w \in \mathbb{C}$. It is clear that an isometry preserves the lengths of curves and therefore maps geodesics into geodesics. It also preserves the area of regions. A fundamental result in differential geometry is the Gauss–Bonnet theorem, which states the existence of a real-valued function $K(z)$, called the curvature of the metric at the point z, such that for every geodesic triangle Δ with internal angles α, β, γ we have

$$\alpha + \beta + \gamma - \pi = \iint_\Delta K \, d\sigma \,,$$

where $d\sigma$ is the area element of the metric (the area of a region R being therefore $A(R) = \iint_R d\sigma$). Finally we say that the metric is *complete* if every geodesic has infinite length.

Theorem 3.1.2 (Hyperbolic plane) *There exists a Riemannian metric on $\mathbb{D}$ whose isometry group coincides with the group of conformal diffeomorphisms of $\mathbb{D}$. Any such metric is complete and conformal and has constant negative curvature. There exists a unique such metric with curvature -1. The geodesics of this metric are the circles orthogonal to the boundary of $\mathbb{D}$.*

Proof The positive multiples of the Euclidean inner product $\langle \cdot, \cdot \rangle$ are the only inner products in $\mathbb{C}$ that are invariant under rotations and reflection with respect to the real axis ($z \mapsto \bar{z}$). Therefore, on the one hand any Riemannian metric on $\mathbb{D}$ that has the conformal group as isometry group must induce an inner product at the origin which is a positive multiple of the Euclidean inner product. On the other hand let $\langle \cdot, \cdot \rangle_0$ be any positive multiple of the Euclidean inner product. As we have already seen, given $z \in \mathbb{D}$ there exists a Möbius transformation $\phi \in \text{Möb}(\mathbb{D})$ such that $\phi(0) = z$ and any other such automorphism is the composition of ϕ with a rotation. We can define an inner product

$\langle \cdot, \cdot \rangle_z$ by $\langle v, w \rangle_z = \langle (\phi^{-1})'(z)v, (\phi^{-1})'(z)w \rangle_0$, and we see that, by the invariance of $\langle \cdot, \cdot \rangle_0$ under rotations, it does not depend on the choice of ϕ. Therefore this defines a Riemannian metric such that all conformal diffeomorphisms of $\mathbb{D}$ are isometries. Since the group of isometries acts transitively, it follows that the curvature of the metric is constant. From the uniqueness theorem for geodesics it follows that any curve that is the set of fixed points of an isometric involution must be a geodesic. Indeed, let f be an isometric involution, i.e. $f \neq$ id is an isometry and $f \circ f(z) = z$, and let C be the curve of fixed points of f. If C is not a geodesic, there exists a geodesic γ different from C which is tangent to C at some point $z \in C$. But then $f(\gamma)$ is different from γ and it is also a geodesic with the same tangent vector, which is a contradiction. Now, on the one hand any circle orthogonal to the boundary of $\mathbb{D}$ is the set of fixed points of the corresponding inversion, which is an isometric involution of our metric. Hence each such circle is a geodesic. On the other hand, given any $z \in \mathbb{D}$ and any vector $v \in \mathbb{C}$, there is a circle $C \in \mathcal{D}_z$ that is tangent to v. Hence these circles comprise all the geodesics. Next, let us consider three distinct points in the boundary of $\mathbb{D}$ and the geodesics connecting each pair. This gives a geodesic triangle with vertices at infinity and internal angles equal to zero. Perturbing the geodesics slightly we get a geodesic triangle with vertices near infinity and internal angles close to zero. Since the curvature of the metric is constant we obtain from the Gauss–Bonnet formula that the curvature must be negative.

So far we have constructed a one-real-parameter set of metrics, all of which have the same geodesics. Now, if we fix a geodesic triangle Δ with internal angles α, β, γ we have that $\alpha + \beta + \gamma - \pi = K(m)A(m)$, where $K(m)$ is the curvature of the metric given by the value m of the parameter and $A(m)$ is the area of the triangle Δ. Since the length of a vector in the metric corresponding to the value m is $\sqrt{m}$ times the length of the same vector in the metric corresponding to $m = 1$, we have that $A(m)$ grows linearly with m. Hence there is a unique value of m such that $K(m) = -1$. $\qquad\qquad\square$

The Riemannian metric constructed above is called the *hyperbolic metric* or Poincaré metric. Since $\mathbb{H}$ is diffeomorphic to $\mathbb{D}$ via a Möbius transformation, there is a unique Riemannian metric in $\mathbb{H}$ for which the above Möbius transformation is an isometry. This is called the Poincaré metric of $\mathbb{H}$. Since this metric is conformal we may write $|v|_z = \rho_{\mathbb{H}}(z)|v|$ where $\rho_{\mathbb{H}}$ is a positive function. Since translations by real numbers are isometries of the hyperbolic metric, $\rho_{\mathbb{H}}(z)$ does not depend on the real part of z. Since multiplications by positive real numbers are also isometries, we deduce that $\rho_{\mathbb{H}}(z)$ is a constant multiple of $1/\operatorname{Im} z$. A simple calculation, using the fact that the curvature is -1, shows that in fact $\rho_{\mathbb{H}}(z) = 1/\operatorname{Im} z$. We know also that the vertical lines are geodesics. Hence, using the expression for the density of the hyperbolic metric, we deduce that the hyperbolic distance between the points ai and bi, $b > a > 0$, is $|\log(b/a)|$. From this we can derive a formula for the hyperbolic distance between any two points. Let us consider the *cross-ratio* of four distinct points in the Riemann sphere, which is given by the expression

$$\operatorname{Cr}(z_1, z_2, z_3, z_4) = \frac{(z_1 - z_4)(z_2 - z_3)}{(z_1 - z_2)(z_3 - z_4)} \, .$$

The reader can easily check that the cross-ratio is preserved by the generators of the Möbius group and, consequently, by any Möbius transformation ϕ; in other words

$$\operatorname{Cr}(\phi(z_1), \phi(z_2), \phi(z_3), \phi(z_4)) = \operatorname{Cr}(z_1, z_2, z_3, z_4) \, .$$

Now, the hyperbolic distance between the points ai and bi that we have just computed is equal to $\log(1 + \operatorname{Cr}(0, ai, bi, \infty))$. Since any geodesic can be mapped to this vertical line by an isometry, we deduce by the invariance of the cross-ratio that the hyperbolic distance between any two points $z, w \in \mathbb{H}$ is equal to

$$\operatorname{dist}_{\mathbb{H}}(z, w) = \log\left(1 + \operatorname{Cr}(z_\infty, z, w, w_\infty)\right)$$

where z_∞, w_∞ are the points at infinity of the geodesic connecting z and w. This formula is also valid in $\mathbb{D}$, since all the ingredients are invariant under isometries.

Example 3.1.3 *The spherical metric.* Let us consider another example of a Riemannian metric of constant curvature and whose isometry group also acts transitively. One can describe the spherical metric using stereographic projection, which is a conformal diffeomorphism between the sphere $S^2 = \{x = (x_1, x_2, x_3) \in \mathbb{R}^3; x_1^2 + x_2^2 + x_3^2 = 1\}$ and $\widehat{\mathbb{C}}$. We embed the complex plane in $\mathbb{R}^3$ by $x_1 + ix_2 \mapsto (x_1, x_2, 0)$ and let $N = (1, 0, 0)$ be the north pole of the sphere. We map N to the point ∞ and every other point p in the sphere to the intersection with the horizontal plane of the line that connects N to p. If we consider in S^3 the Riemannian metric induced by the Euclidean metric of $\mathbb{R}^3$, we can verify that the map just defined is conformal. Therefore there is a unique conformal Riemannian metric on $\widehat{\mathbb{C}}$ that makes the stereographic projection an isometry. This is the spherical metric. The stereographic projection conjugates the group of rotations with a subgroup of the group of Möbius transformations. Since the group of rotations of the sphere is contained in the group of isometries and acts transitively, the same happens with the above-mentioned subgroup of Möb($\widehat{\mathbb{C}}$). In fact, one can easily see that this subgroup is precisely the set of Möbius transformations of the form

$$z \mapsto \frac{az - \overline{b}}{bz + \overline{a}} \, ,$$

where a and b are complex numbers satisfying $|a|^2 + |b|^2 = 1$. In particular, taking $a = 0$ and $b = i$ we see that $z \mapsto 1/z$ is an isometry of the spherical metric. Finally, the spherical norm of a vector $v \in \mathbb{C}$ at a finite point z is given by $\|v\|_z = \rho_{\widehat{\mathbb{C}}}(z)|v|$, where

$$\rho_{\widehat{\mathbb{C}}}(z) = \frac{2}{1 + |z|^2} \, .$$

We finish this section by stating another version of Schwarz's lemma.

Lemma 3.1.4 (Schwarz lemma) *If $f\colon \mathbb{D} \to \mathbb{D}$ is a holomorphic map then either f strictly contracts the hyperbolic metric or it is an isometry.*

Proof Suppose f is not an isometry. Let $z \in \mathbb{D}$ and $v \in \mathbb{C}$, $v \neq 0$. We must prove that $|f'(z)v|_w < |v|_z$ where $w = f(z)$. If $z = w = 0$ then this is just the classical Schwarz lemma. Otherwise, let ϕ and ψ be automorphisms of $\mathbb{D}$ such that $\phi(0) = z$ and $\psi(w) = 0$. Now, again apply the classical Schwarz lemma to $\psi \circ f \circ \phi$ and use the fact that ϕ and ψ are isometries. $\square$

3.2 Some topological results

A subset $S \subseteq \widehat{\mathbb{C}}$ is *connected* if, for any pair of disjoint open sets A, B with $S = (A \cap S) \cup (B \cap S)$, either $A \cap S = \emptyset$ or $B \cap S = \emptyset$; S is *pathwise connected* if any two points of S can be connected by a continuous curve entirely contained in S. We say that S is *locally connected* (*pathwise locally connected*) if every point of S has arbitrarily small connected (pathwise connected) neighborhoods. It is clear that an open set is connected if and only if it is pathwise connected. The *connected component* of a point $x \in S$ is the largest connected subset of S that contains x.

Two continuous curves $\gamma_0, \gamma_1 \colon [0,1] \to U$ are *homotopic* (relative to their end-points) if there exists a continuous map $h \colon [0,1] \times [0,1] \to U$ such that $h(t,0) = \gamma_0(t)$, $h(t,1) = \gamma_1(t)$, $h(0,s) = \gamma_0(0)$ and $h(1,s) = \gamma_0(1)$ for all $(t,s) \in [0,1] \times [0,1]$. The map h is called a *homotopy* between γ_0 and γ_1.

A subset of the Riemann sphere is *simply connected* if any closed curve in it is homotopic to a constant curve. It follows that an open subset of the Riemann sphere is simply connected if and only if its complement is connected. Hence the simply connected open subsets of the Riemann sphere are precisely the complements of the connected compact subsets.

A *covering map* is a local homeomorphism $\pi \colon U \to V$ such that every point in V has a neighborhood W with the following property: the restriction of π to each connected component of $\pi^{-1}(W)$ is a homeomorphism onto W. A *covering automorphism* or deck transformation is a homeomorphism $f \colon U \to U$ such that $\pi \circ f = \pi$. It is clear that the set of automorphisms of a covering map is a subgroup of the group of homeomorphisms of U. If the covering map is holomorphic then the automorphisms are holomorphic diffeomorphisms.

Example 3.2.1 *Holomorphic covering of* $\mathbb{C} \setminus \{0\}$. The exponential map $\exp z = e^{\operatorname{Re} z}(\cos \operatorname{Im} z + i \sin \operatorname{Im} z)$ is a holomorphic covering map from $\mathbb{C}$ to $\mathbb{C} \setminus \{0\}$ whose automorphism group is the group of translations $\{z \mapsto z + 2k\pi i : k \in \mathbb{Z}\}$.

Example 3.2.2 *Holomorphic covering of* $\mathbb{D} \setminus \{0\}$. The map $\Phi \colon \mathbb{H} \to \mathbb{D} \setminus \{0\}$ defined by $\Phi(z) = \exp(2\pi i z)$ is a holomorphic covering map whose automorphism group is the translation group $\{z \mapsto z + k : k \in \mathbb{Z}\}$.

Example 3.2.3 *Holomorphic coverings of annuli* (figure 3.1). Consider the annulus $A_R = \{z \in \mathbb{C} : 1 < |z| < R\}$. The map $\Psi \colon \mathbb{H} \to A_R$ defined by

$$\Psi(z) = \exp\left\{-\frac{2\pi i}{\log \lambda} \operatorname{Log} z\right\},$$

where $\lambda = e^{\pi^2/\log R}$ and $\operatorname{Log} re^{i\theta} = \log r + i\theta$, is a holomorphic covering map whose automorphism group is the group of homotheties $\{z \mapsto \lambda^k z : k \in \mathbb{Z}\}$.

Example 3.2.4 Let $f \colon \widehat{\mathbb{C}} \to \widehat{\mathbb{C}}$ be a holomorphic map of degree $d \geq 2$ (figure 3.2). Let $C(f) = \{z \in \widehat{\mathbb{C}} : f'(z) = 0\}$ be the set of critical points of f. Consider the open sets $V = \widehat{\mathbb{C}} \setminus f(C(f))$ and $U = \widehat{\mathbb{C}} \setminus f^{-1}(f(C(f)))$.

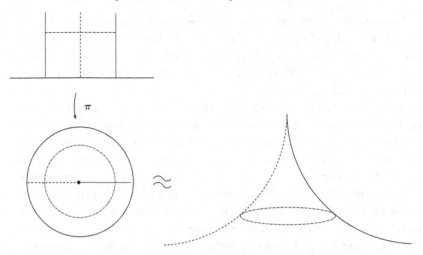

Fig. 3.1. Holomorphic covering of a parabolic annulus.

Then the restriction of f to U is a holomorphic covering map such that the fiber over each point has exactly d elements. In most of these examples, the automorphism group reduces to the identity because each automorphism must be the restriction of a Möbius transformation. The special case $z \mapsto z^d$ is a covering map from $\mathbb{C} \setminus \{0\}$ onto itself whose automorphism group is the group of rotations $\{z \mapsto e^{2\pi ki/d}z \ : \ 0 \le k < d\}$.

Let $\pi \colon U \to V$ be a covering map. Let $f \colon X \to V$ be a continuous map. A *lift* of f is a continuous map $\hat{f} \colon X \to U$ such that $\pi \circ \hat{f} = f$. Notice that if $\hat{f}$ is a lift of f and ϕ is an automorphism of the covering then $\phi \circ \hat{f}$ is also a lift of f. However, if $\hat{f}_1$ and $\hat{f}_2$ are lifts of f then the set of points where $\hat{f}_1$ coincides with $\hat{f}_2$ is both open and closed. Therefore, if X is connected then two lifts that coincide at one point are identical. The existence of a lift is a much more delicate problem that has a positive answer if X is simply connected.

Theorem 3.2.5 *Let $\pi \colon U \to V$ be a covering map.*

(i) *If $f \colon [0,1] \to V$ is a continuous curve and $\pi(z_0) = f(0)$ then there exists a unique lift $\hat{f} \colon [0,1] \to U$ of f satisfying $f(0) = z_0$.*

(ii) *If $h \colon [0,1] \times [0,1] \to V$ is a homotopy and $\pi(z_0) = h(0,0)$ then there exists a unique lift $\hat{h} \colon [0,1] \times [0,1] \to U$ of h satisfying $\hat{h}(0,0) = z_0$.*

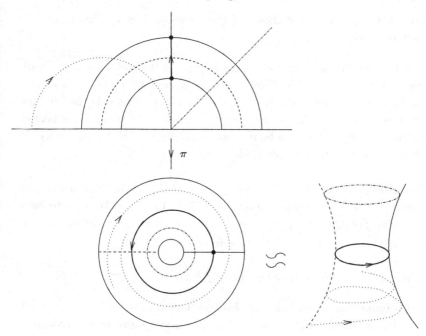

Fig. 3.2. Holomorphic covering of a hyperbolic annulus.

(iii) *If X is simply connected, $f\colon X \to V$ is continuous and $\pi(z_0) = f(x_0)$ then there exists a unique lift $\hat{f}\colon X \to U$ such that $f(x_0) = z_0$.*

Proof We have already discussed the uniqueness of the lift. To prove its existence in the first case, let us consider the set A of all $t \in [0,1]$ such that the restriction of f to $[0,t]$ has a lift starting at z_0. It is clear that A contains a neighborhood of 0 and that A is open and closed in $[0,1]$. Since $[0,1]$ is connected we have that $A = [0,1]$. The second case is proved similarly, by considering the set A of values (t,s) such that the restriction of h to $[0,t] \times [0,s]$ has a lift starting at z_0. Now let us consider the third case. Given $x \in X$ let us choose a curve $\gamma\colon [0,1] \to X$ such that $\gamma(0) = x_0, \gamma(1) = x$. By the first part of the lemma, the curve $f \circ \gamma$ has a lift starting at z_0. We define $\hat{f}(x)$ as the endpoint of this lift. This does not depend on the choice of the curve γ because if $\tilde{\gamma}$ is another such curve then it is homotopic to γ because X is simply connected. Thus $f \circ \gamma$ is homotopic to $f \circ \tilde{\gamma}$ and by the second part of the lemma we can lift this homotopy. In particular it follows that the two lift curves have the same endpoints, proving that the construction does not depend on the choice of the curve γ. The continuity of $\hat{f}$ is easy to check.   □

An important special case occurs when f is the inclusion of a simply connected subset $X \subset V$. For each $z_0 \in U$ such that $\pi(z_0) = x_0$ there exists a continuous function $\phi\colon X \to U$ such that $\pi \circ \phi$ is the identity of X and $\phi(x_0) = z_0$. Hence the restriction of π to each connected component of $\pi^{-1}(X)$ is a homeomorphism onto X. If π is holomorphic then clearly ϕ is also holomorphic.

Corollary 3.2.6 *Let $\pi_i\colon U_i \to V_i$, $i = 1, 2$, be covering maps and suppose that U_1 is simply connected. If $f\colon V_1 \to V_2$ is a continuous map and $f(\pi_1(z_1)) = \pi_2(z_2)$ then there exists a unique continuous map $\hat{f}\colon U_1 \to U_2$ such that $\pi_2 \circ \hat{f} = f \circ \pi_1$ and $\hat{f}(z_1) = z_2$.*

Proof It is enough to take the lift of $f \circ \pi_1$.   □

An important special case of this corollary occurs when $\pi_1 = \pi_2$ and f is the identity. In this case $\hat{f}$ is an automorphism of the covering. We deduce that when U is simply connected the automorphism group acts transitively on each fiber: if $\pi(z_1) = \pi(z_2)$ then there exists an automorphism that maps z_1 to z_2.

Topological obstructions to the lifting of maps exist whose domains are not simply connected. These obstructions can be formulated in terms of the fundamental group. Let us discuss briefly the main concepts involved. Let X be a topological space that is connected and locally connected, and let x_0 be a point of X. The *fundamental group* of X with base point x_0 is the set $\pi_1(X, x_0)$ of homotopy classes of closed curves with endpoints at x_0. The group multiplication is defined as follows: the product of the homotopy class of a curve $\gamma_1\colon [0, 1] \to X$ and the homotopy class of a curve $\gamma_2\colon [0, 1] \to X$ is the homotopy class of the curve $\gamma\colon [0, 1] \to X$ defined by

$$\gamma(t) = \begin{cases} \gamma_2(2t) & \text{if } 0 \le t \le \tfrac{1}{2}, \\[2mm] \gamma_1(2t - 1) & \text{if } \tfrac{1}{2} \le t \le 1. \end{cases}$$

One can prove that the homotopy class of γ does not depend on the choice of γ_1 and γ_2 but only on their homotopy classes. Also, this multiplication satisfies the group axioms (in general it is non-commutative). A continuous map $f\colon X \to Y$, with $f(x_0) = y_0$, induces a group homomorphism $f_*\colon \pi_1(X, x_0) \to \pi_1(Y, y_0)$, i.e. the image of the homotopy class of a closed curve γ is the homotopy class of the curve $f \circ \gamma$. One can prove that this is well defined and that if two maps are homotopic relative to the base points then they induce the same homomorphism of the fundamental groups. We can now state a general existence theorem for lifts that includes the previous theorem as a special case. We refer to [Mas] for its proof.

Theorem 3.2.7 *Let $\pi\colon U \to V$ be a covering map and $f\colon X \to V$ a continuous map. We assume that all the spaces involved are locally connected. Let $f(x_0) = z_0 = \pi(w_0)$. Then there exists a lift $\hat{f}\colon X \to U$ with $\hat{f}(x_0) = w_0$ if and only if $f_*(\pi_1(X, x_0)) \subset \pi_*(\pi_1(U, w_0))$.* $\square$

Corollary 3.2.8 *Let $\pi\colon U \to V$ be a covering map. Let $f_i\colon X \to V$, $i = 1, 2$, be continuous maps such that $f_1(x_0) = f_2(x_0) = z_0 = \pi(w_0)$. Suppose that f_1 is homotopic to f_2 relative to x_0. Then if f_1 has a lift mapping x_0 into w_0 so does f_2.* $\square$

We finish this section by discussing an existence theorem for covering spaces in the special case we will need. Let V be an open subset of the Riemann sphere and $z_0 \in V$. Let $\tilde{V}$ be the set of homotopy classes of continuous curves with initial point z_0. Let $\pi\colon \tilde{V} \to V$ be the map that associates with the homotopy class of a curve γ the endpoint of γ (notice that all curves in the homotopy class of γ have the same initial point and the same endpoint). Next we define a topology in $\tilde{V}$ by describing a system of neighborhoods of each point. So let us consider $[\gamma]$, the homotopy class of a curve γ. Let z_1 be the endpoint of γ. Let Δ be a disk in V with center z_1. Let $\tilde{\Delta}([\gamma])$ be the set of homotopy classes of the curves that start at z_0, follows γ up to z_1 and then continue until the endpoint in Δ, following a straight segment in Δ. Clearly π is a bijection between $\tilde{\Delta}([\gamma])$ and Δ. Now we define the topology of $\tilde{V}$ by declaring each $\tilde{\Delta}([\gamma])$ to be an open neighborhood. We have the following result.

Theorem 3.2.9 *The map $\pi\colon \tilde{V} \to V$ is a covering map and $\tilde{V}$ is simply connected.*

The reader could prove this theorem as an exercise using the topology defined above, or look the proof up in [Mas].

3.3 Hyperbolic contraction

Let $\pi\colon U \to V$ be a smooth covering map, i.e. π is also a C^∞ local diffeomorphism. Any Riemannian metric on V can be lifted to a unique Riemannian metric on U so that π becomes a local isometry. Indeed, it is enough to define $\langle v, w \rangle_z = \langle D\pi(z)v, D\pi(z)w \rangle_{\pi(z)}$. All covering automorphisms are isometries of this metric. Conversely, if U is simply connected and we start with a Riemannian metric on U such that the covering automorphisms are isometries, then there exists a unique Riemannian metric on V such that π is a local isometry. To define the inner product of two vectors v, w at a point $z \in V$ we take any point $\tilde{z}$ in the fiber of z and define $\langle v, w \rangle_z = \langle (D\pi(z))^{-1}v, (D\pi(z))^{-1}w \rangle_{\tilde{z}}$. This does not depend on the choice of $\tilde{z}$ because the automorphism group acts transitively on the fiber and, by hypothesis, it is contained in the isometry group.

Let us consider a holomorphic covering map $\pi\colon \mathbb{D} \to V$. Since the covering automorphism group is a subgroup of the group of Möbius transformations, which is a subgroup of the isometry group of the hyperbolic metric of $\mathbb{D}$, it follows that there exists a unique complete Riemannian metric on V that is locally isometric to the hyperbolic metric on $\mathbb{D}$. This is the *hyperbolic* metric of V. We will show in section 3.6 that any connected open subset of the Riemann sphere whose complement contains at least three points has a hyperbolic metric.

Theorem 3.3.1 (Schwarz lemma) *Let $f\colon V_1 \to V_2$ be a holomorphic map between open sets endowed with a hyperbolic metric. Then either the derivative of f strictly contracts the hyperbolic metric at every point or f is a local isometry and also a covering map.*

Proof Let $\hat{f}\colon \mathbb{D} \to \mathbb{D}$ be a lift of f. It is enough to apply the previous version of the Schwarz lemma to $\hat{f}$.   □

3.3.1 Markov maps

Let us present a first dynamical application of the Schwarz lemma. The setup here will be generalized later (see section 3.9).

Definition 3.3.2 *A Markov map is a holomorphic map $f\colon U \to V$ with the following properties:*

(i) *the open set $V \subset \mathbb{C}$ is a topological disk;*

(ii) *the open set U is a finite union of topological disks whose closures are pairwise disjoint and are contained in V;*

(iii) *the restriction of f to each component of U is a univalent map onto V.*

The typical situation for $d = 2$ is shown in figure 3.3. The application we have in mind is the following. See also the more general result given in theorem 3.9.4.

Theorem 3.3.3 *Let $f\colon U \to V$ be a Markov map. Then $J_f = \{z \in U : f^n(z) \in U \text{ for all } n \in \mathbb{N}\}$ is a compact f-invariant set. The restriction of f to J_f is topologically conjugate to the full one-sided shift map*

$$\sigma\colon \{1, 2, \ldots, d\}^{\mathbb{N}} \to \{1, 2, \ldots, d\}^{\mathbb{N}}$$

defined by $\sigma((x_n)_{n \geq 0}) = (x_{n+1})_{n \geq 0}$, where d is the number of components of U.

Proof Let $U_1, \ldots, U_d$ be the connected components of U. On the one hand, Schwarz's lemma implies that the inclusion of each U_i in V is a strict contraction of the respective hyperbolic metrics. On the other hand, since the restriction of f to each U_i is a univalent map onto V,

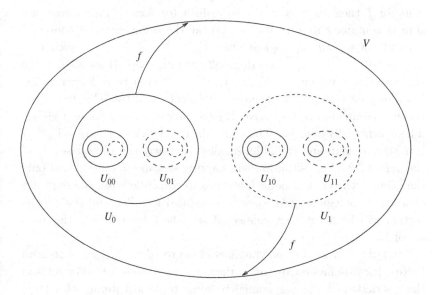

Fig. 3.3. A Markov map with $d = 2$.

it is an isometry of the hyperbolic metrics. Therefore the derivative of f strictly expands the hyperbolic metric of V at each point of U, i.e. there exists $\lambda > 1$ such that $\|Df(z)v\|_{f(z)} > \lambda\|v\|_z$ for all $z \in U$ and $v \in \mathbb{C}$. For each finite sequence $a_0, a_1, \ldots, a_n$ of elements of $\{1, 2, \ldots, d\}$, let $U_{a_0, a_1, \ldots, a_n}$ be the set of points $z \in U_{a_0}$ such that $f^i(z) \in U_{a_i}$ for $i \leq n$. Since $f(U_{a_0, a_1, \ldots, a_n}) = U_{a_1, \ldots, a_n}$ we have that the hyperbolic diameter of $U_{a_0, a_1, \ldots, a_n}$ is smaller than λ^{-n} times the hyperbolic diameter of U_{a_n}. Since U is compactly contained in V, the hyperbolic metric of V in U is equivalent to the Euclidean metric. Hence the Euclidean diameter of these sets also decays exponentially with n. Furthermore, $U_{a_0, \ldots, a_n, \ldots, a_{n+1}} \subset U_{a_0, a_1, \ldots, a_n}$. Therefore, given an infinite sequence $a_0, a_1, \ldots$ there exists a unique point in the intersection of all the disks $U_{a_0, \ldots, a_n}$. This defines a map h from $\{1, 2, \ldots, d\}^{\mathbb{N}}$ into J_f. It is easy to see that h is a homeomorphism onto J_f and conjugates the shift σ to our map f. $\qquad\square$

Example 3.3.4 Here is an example of a Markov map that appears naturally in the iteration of polynomials. Let f be a quadratic polynomial of the form $f(z) = z^2 + c$. If V_0 is a disk centered at the origin with sufficiently large radius then $V_1 = f^{-1}(V_0)$ is a topological disk whose closure is contained in V_0. Now if V_1 contains the critical value of f then $V_2 = f^{-1}(V_1)$ is again a topological disk whose closure is contained in V_1. We can repeat the argument and construct a family of nested topological disks $V_n \subset \overline{V}_n \subset V_{n-1} \ldots$ such that $f: V_j \setminus \{0\} \to V_{j-1} \setminus \{c\}$ is a degree-2 covering map. If for some n the critical value c belongs to $V_{n-1} \setminus V_n$ then $f^{-1}(V_n)$ is no longer a disk but a disjoint union of two topological disks U_0 and U_1. These are compactly contained in U_n and have disjoint interiors, and f maps each one diffeomorphically onto V_n. Therefore, the restriction of f to $U = U_0 \cup U_1$ is a Markov map and the Julia set of f is clearly the Cantor set J_f of theorem 3.3.3. This situation will happen at some n value if and only the orbit of the critical point 0 escapes to ∞. Otherwise, the sequence of nested topological disks V_n will be defined for all n and their intersection will be a compact connected set whose boundary is the Julia set of f.

Similarly, considering polynomials of degree d we deduce that if all finite critical points escape to ∞ then the Julia set is a Cantor set and the restriction of the polynomial to some backward iterate of a large disk is a Markov map, this time with d components.

3.4 Uniformization of bounded simply connected domains

In this section we will show that any *bounded* simply connected open subset of the plane is holomorphically equivalent to the unit disk. Later we will extend this result to all simply connected open subsets of the plane not equal to the plane itself.

Proposition 3.4.1 *Given $v \in \mathbb{D}$, there exists a holomorphic map $f \colon \mathbb{D} \to \mathbb{D}$ with a unique critical point $c \in \mathbb{D}$ such that $f \colon \mathbb{D} \setminus \{c\} \to \mathbb{D} \setminus \{v\}$ is a degree-2 covering map and $f(0) = 0$.*

Proof Let $q(z) = z^2$. Let $\phi \colon \mathbb{D} \to \mathbb{D}$ be a Möbius transformation that maps 0 to v. Let $g = \phi \circ q \circ \phi^{-1}$; note that $g(v) = v$. Let $w \in \mathbb{D}$ such that $g(w) = 0$ and let $\psi \colon \mathbb{D} \to \mathbb{D}$ be a Möbius transformation that maps 0 to w. Take $f = g \circ \psi$. This map fixes 0 and has a unique critical point at $c = \psi^{-1}(v)$ with critical value v. $\qquad\square$

Corollary 3.4.2 *Let $U \subset \mathbb{D}$ be a simply connected open subset that contains 0. If $U \neq \mathbb{D}$ then there exists a univalent function $F \colon U \to \mathbb{D}$ such that $F(0) = 0$ and $|F'(0)| > 1$.*

Proof Let $v \in \mathbb{D} \setminus U$ and $f \colon \mathbb{D} \to \mathbb{D}$, as in the proposition above. On the one hand, since f has a critical point, Schwarz's lemma implies that $|f'(0)| < 1$. On the other hand, since $f \colon \mathbb{D} \setminus \{c\} \to \mathbb{D} \setminus \{v\}$ is a covering map and $U \subset \mathbb{D} \setminus \{v\}$ is simply connected there exists a univalent map $F \colon U \to \mathbb{D}$ such that $F(0) = 0$ and $f \circ F$ is the identity of U. Since $F'(0)f'(0) = 1$, we have $|F'(0)| > 1$. $\qquad\square$

Theorem 3.4.3 *Let $W \subset \mathbb{C}$ be a bounded simply connected open set and let $z_0 \in W$. Then there exists a holomorphic diffeomorphism $\phi \colon W \to \mathbb{D}$ such that $\phi(z_0) = 0$. Any other such diffeomorphism is the composition of ϕ with a rotation.*

Proof Consider the family $\mathcal{F}$ of all univalent maps from W into $\mathbb{D}$ mapping z_0 into 0. Since W is bounded, this family is clearly non-empty: the composition of a translation with a linear contraction serves as an example. Since $\mathcal{F}$ is a uniformly bounded family of holomorphic maps it is equicontinuous and hence a normal family by Montel's theorem. In particular the function $f \in \mathcal{F} \mapsto |f'(0)|$ is bounded. Let $f_n \in \mathcal{F}$ be a sequence of functions such that $|f_n'(0)|$ converges to $\sup\{|f'(0)| : f \in \mathcal{F}\}$. Since $\mathcal{F}$ is a normal family, we can assume, by passing to a subsequence, that f_n converges uniformly in compact subsets of W to a map ϕ which is holomorphic. Since the derivative of f_n also converges uniformly on compact subsets to the derivative of ϕ we have that $\phi'(z_0) = \sup\{|f'(0)| : f \in \mathcal{F}\}$. In particular ϕ is not constant and, therefore, must be univalent and also its image $U = \phi(W)$ must be a simply connected subset of $\mathbb{D}$. However, if $U \neq \mathbb{D}$ then the above corollary implies the existence of a univalent map $F : U \to \mathbb{D}$ such that $F(0) = 0$ and $|F'(0)| > 1$. This is not possible because $F \circ \phi$ would be in $\mathcal{F}$ and the absolute value of its derivative at 0 would be greater than $\sup\{|f'(0)| : f \in \mathcal{F}\}$. This contradiction proves the theorem. $\qquad\square$

The inverse of the holomorphic diffeomorphism ϕ constructed in the proof of the theorem 3.4.3 is called the *Riemann mapping* of the simply connected open set W. The behavior of the Riemann mapping at the boundary of the disk depends very much on the topological properties of the boundary of W. For the proof of the following theorem, see [Mar, p. 70].

Theorem 3.4.4 *Let W be a simply connected domain of the Riemann sphere whose complement contains at least two points. Then*

   (i) *the Riemann mapping of W extends continuously to the boundary of $\mathbb{D}$ if and only if ∂W is locally connected;*

  (ii) *the Riemann mapping extends to a homeomorphism of the closure of $\mathbb{D}$ onto the closure of W if and only if ∂W is a Jordan curve.* $\qquad\square$

The Riemann mapping also depends continuously on the Jordan domain if the former is conveniently normalized. The theorem below was proved by Markushevich [Mar, theorem 2.6, p. 72], and O. J. Farrel [Far, theorem III, p. 373].

Theorem 3.4.5 *Let $V_1 \supset V_2 \supset \cdots$ be a decreasing sequence of uniformly bounded Jordan domains whose intersection is a Jordan domain V. Let $z \in V$, and let $\phi_n : V_n \to \mathbb{D}$ be the Riemann mapping normalized in such a way that $\phi_n(z) = 0$ and the derivative of ϕ_n at z is a positive real number. Then the restrictions of ϕ_n to the closure $\overline{V}$ of V converge uniformly to the Riemann mapping of V.* $\qquad\square$

3.5 The holomorphic universal covering of $\mathbb{C} \setminus \{0,1\}$

In this section we will prove the existence of a hyperbolic metric in the complement of three points in the Riemann sphere and study properties of this metric.

Theorem 3.5.1 *There exists a holomorphic covering map* $\pi \colon \mathbb{D} \to \mathbb{C} \setminus \{0,1\}$.

Proof Let $T^+ \subset \mathbb{D}$ be the geodesic ideal triangle with vertices $-1, 1, i$. By theorem 3.4.4 there exists a homeomorphism ϕ^+ from the closure of T^+ onto the closure of $\mathbb{H}$ that is a holomorphic diffeomorphism from T^+ onto $\mathbb{H}$. Taking the composition with a suitable Möbius transformation, we may assume that ϕ^+ maps -1 to $0, 1$ to 1 and i to ∞. Using the Schwarz reflection principle [Rud, theorem 11.17], we can extend ϕ^+ to a holomorphic diffeomorphism ϕ^- of the ideal triangle T^- obtained from T^+ by reflection with respect to the side $(-1,1)$ onto the lower half-space $\mathbb{H}^-$. We define π to be ϕ^+ on T^+ and ϕ^- on T^-, and the Schwarz reflection principle asserts that π will be holomorphic not only in the union of those triangles but also at the common side. Similarly we extend ϕ to new triangles by using reflections with respect to the other sides. The two new triangles are also mapped onto the lower half-space. Now using reflection with respect to the sides of the new triangles, we extend π to adjacent triangles by mapping them in the upper half-space. Continuing this process indefinitely, we get a triangulation of $\mathbb{D}$ and a holomorphic map π that maps each triangle onto either the upper half-space or the lower half-space and each side of a triangle onto a segment of the real line: $(-\infty, 0)$, $(0,1)$ or $(1, \infty)$. This is clearly a holomorphic covering of $\mathbb{C} \setminus \{0,1\}$; see figure 3.4. □

We have just proved the existence of a hyperbolic metric on the sphere minus three points, i.e. $\widehat{\mathbb{C}} \setminus \{0, 1, \infty\}$. Next we want to compare this conformal metric with the spherical metric.

Theorem 3.5.2 *Consider the spherical metric*

$$|ds| = \frac{2|dz|}{1 + |z|^2} \,,$$

and let $d \colon \widehat{\mathbb{C}} \times \widehat{\mathbb{C}} \to \mathbb{R}^+$ *be the corresponding distance function. Let* $\rho_0 |ds|$ *be the hyperbolic metric of* $S = \widehat{\mathbb{C}} \setminus \{0, 1, \infty\}$. *Then there exist positive*

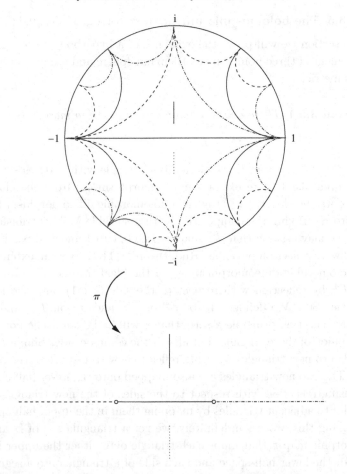

Fig. 3.4. Holomorphic covering of $\mathbb{C} \setminus \{0, 1\}$. The upper figure shows geodesic lines of three types (represented by broken lines, solid lines and broken-and-dotted lines). These are mapped respectively onto the broken, solid and broken-and-dotted lines in the lower figure.

constants C_1, C_2 such that

$$\frac{C_1}{d(w, \partial S)|\log d(w, \partial S)|} \leq \rho_0(w) \leq \frac{C_2}{d(w, \partial S)|\log d(w, \partial S)|} ,$$

where $\partial S = \{0, 1, \infty\}$.

Proof We say that two positive functions ρ_1, ρ_2 are commensurable, $\rho_1 \asymp \rho_2$, if there exist positive constants C_1 and C_2 such that $C_1\rho_1 \leq$

$\rho_2 \leq C_2 \rho_1$. So we want to prove that the function ρ_0 is commensurable with the function $\rho_1(w) = 1/d(w, \partial S) |\log d(w, \partial S)|$.

Let $\pi \colon \mathbb{D} \to \mathbb{C} \setminus \{0, 1\}$ be the covering map of theorem 3.5.1. Let B_j, $j = -1, 1, i$, be disjoint horoballs with vertices at the points $-1, 1, i$, i.e. $B_j \subset \mathbb{D}$ is an Euclidean disk tangent to the boundary of $\mathbb{D}$ at the point j. We notice that each horoball B_j is invariant under the cyclic subgroup of the group of deck transformations that fixes the vertex of the horoball and that the restriction of π to each horoball is a covering map onto a topological disk D_j of the Riemann sphere, where D_{-1} contains 0, D_1 contains 1 and D_i contains ∞. The complement of the union of these three topological disks is a compact region of S where the two functions are commensurable. It remains to prove that they are commensurable in each D_j. Let $\phi_j \colon \mathbb{D} \to D_j$ be the Riemann mapping that takes 0 to 0 if $j = -1$, to 1 if $j = 1$ and to ∞ if $j = i$. Let us prove that the two functions are commensurable in D_{-1}.

Let $Q \subset \mathbb{D}$ be the geodesic quadrilateral with vertices $-1, -i, 1, i$, which is mapped by π onto $\mathbb{C} \setminus \{0, 1\}$. Since $\hat{\phi}_0$ is a Möbius transformation, we have that the spherical distance $d(\hat{z}, \infty)$ is commensurable with $d(\hat{w}, -1)$ where $\hat{w} = \hat{\phi}_{-1}(\hat{z})$. Similarly, $d(z, 0)$ is commensurable with $d(w, 0)$ where $w = \phi_0(z)$. However, $d(\hat{z}, \infty) \asymp |\log d(z, 0)|^{-1}$ and $\|D\Psi(\hat{z})\|_{\mathrm{sph}} \asymp d(z, 0)$, where $\|D\Psi\|_{\mathrm{sph}}$ is the norm of the derivative in the spherical metric. Furthermore the norms of the derivatives of ϕ_0 and $\hat{\phi}_0$ and of their inverses are bounded in the spherical metric and we have $d(\hat{w}, -1) \asymp |\log d(w, 0)|^{-1}$ and $\|D\pi(\hat{w})\|_{\mathrm{sph}} \asymp d(w, 0)$. Finally, since π is a local isometry in the hyperbolic metric, $\rho_0(w) = \rho_{\mathbb{D}}(\hat{w}) \|D\pi(\hat{w})\|_{\mathrm{sph}}$. This proves that the two functions are commensurable in D_{-1}, since $\rho_{\mathbb{D}}(\hat{w}) \asymp 1/d(\hat{w}, -1)$. Similarly, we can prove that the two functions are commensurable in D_1 and in D_i. $\square$

Corollary 3.5.3 *Given $\epsilon > 0$, there exists a positive constant $C(\epsilon)$ with the following property. Let $U \subset \widehat{\mathbb{C}}$ be an open subset of the Riemann sphere whose complement contains at least three points and which is such that, given $z_1 \in \partial U$, there exist $z_2, z_3 \in \partial U$ with $d(z_i, z_j) \geq \epsilon$. If $\rho_U |ds|$ is the hyperbolic metric of U then for all $w \in U$ we have*

$$\frac{C(\epsilon)}{d(w, \partial U) |\log d(w, \partial U)|} \leq \rho_U(w) \leq \frac{1}{d(w, \partial U)}$$

Proof Let $z_1 \in \partial U$ be such that $d(w, \partial U) = d(w, z_1)$. Let $z_2, z_3 \in \partial U$ as in the statement (for this choice of z_1) and let L be the Möbius transformation that maps z_1 to 0, z_2 to 1 and z_3 to ∞. If the hyperbolic metric of $\widehat{\mathbb{C}} \setminus \{z_1, z_2, z_3\}$ is $\rho|ds|$ then, by Schwarz's lemma, we have on the one hand that $\rho(w) \leq \rho_U(w)$. On the other hand, L is an isometry between the hyperbolic metric of $\widehat{\mathbb{C}} \setminus \{z_1, z_2, z_3\}$ and that of $S = \widehat{\mathbb{C}} \setminus \{0, 1, \infty\}$. Also, with respect to the spherical metric, both L and its inverse are Lipschitz with constants bounded by a function of ϵ. Finally, we can use the left-hand inequality of theorem 3.5.2 to finish the proof of the left-hand inequality in the corollary. The right-hand inequality follows directly from the Schwarz lemma, since the disk in the spherical metric with center w and radius $d(w, \partial U)$ is contained in U, and the density of the hyperbolic metric of this disk at w is given by the right-hand side of the inequality.   □

In particular, the hyperbolic metric of any region U of the Riemann sphere is the product of the spherical metric and a positive function that tends to infinity at the boundary of U.

Lemma 3.5.4 *The family of all holomorphic maps from $\mathbb{D}$ to $\mathbb{C} \setminus \{0, 1\}$ is a normal family.*

Proof Let $f_n \colon \mathbb{D} \to \mathbb{C} \setminus \{0, 1\}$ be a family of holomorphic maps. Passing to a subsequence if necessary, we may assume that $z_n = f^n(0)$ converges to some point z in the Riemann sphere. By the Schwarz lemma, the image produced by f_n of the hyperbolic disk with center 0 and radius r is contained in the hyperbolic disk with center z_n and radius r. If z_n tends to zero then the spherical diameter of this disk must tend to zero. Hence f_n converges uniformly in compact subsets to zero. The same happens if $z = 1$ or $z = \infty$. Next we assume that $z \in \widehat{\mathbb{C}} \setminus \{0, 1, \infty\}$. Let $\hat{z} \in \mathbb{D}$ be such that $\pi(\hat{z}) = z$ where $\pi : \mathbb{D} \to \mathbb{C} \setminus \{0, 1\}$ is the covering map of theorem 3.5.1. Now take a sequence $\hat{z}_n \to \hat{z}$ with $\pi(\hat{z}_n) = z_n$ and let $\hat{f}_n$ be the lift of f_n with $\hat{f}_n(0) = \hat{z}_n$. Since $\hat{f}_n$ is a sequence of bounded holomorphic functions, we have from the classical Montel theorem 2.3.2 that there is a subsequence $\hat{f}_{n_i}$ that converges uniformly on compact subsets to a holomorphic map $\hat{f}$. Clearly $\hat{f}(0) = \hat{z}$ and, therefore, $\hat{f}$ maps $\mathbb{D}$ into $\mathbb{D}$. The composition $f = \pi \circ \hat{f}$ is the limit of the sequence f_{n_i}.   □

Theorem 3.5.5 (Montel) *Let $\epsilon > 0$ be a given number, and let $\mathcal{F}$ be a family of holomorphic maps from a domain U of the Riemann sphere into $\widehat{\mathbb{C}}$. If for each $f \in \mathcal{F}$ there exist points $z_1(f), z_2(f), z_3(f) \in \widehat{\mathbb{C}} \setminus f(U)$*

such that the spherical distance between any two of them is greater than ϵ *then* $\mathcal{F}$ *is a normal family.*

Proof We may assume that $U = \mathbb{D}$. Let f_n be a sequence of functions in $\mathcal{F}$. For each n let L_n be the Möbius transformation that maps $z_1(f_n)$ to $0, z_2(f_n)$ and 1 and $z_3(f_n)$ to ∞. Passing to a subsequence if necessary, we may assume that $z_j(f_n)$ converges to z_j. It follows that the spherical distance between z_j and z_k is at least ϵ if $j \neq k$. In particular the three points are distinct and L_n converges uniformly to the Möbius transformation L such that $L(z_1) = 0$, $L(z_2) = 1$, $L(z_3) = \infty$. Furthermore, the family $L_n \circ f_n$ satisfies the hypothesis of lemma 3.5.4. Hence, again passing to a subsequence, we may assume that $L_n \circ f_n$ converges uniformly on compact subsets to a map g. Thus f_n converges to $L^{-1} \circ g$ uniformly on compact subsets. $\qquad\square$

Corollary 3.5.6 (Compactness) *Consider the family* $\mathcal{U}$ *of all univalent functions from* $\mathbb{D}$ *to* $\mathbb{C}$ *that fix the origin and whose derivative at* 0 *is equal to* 1. *Then* $\mathcal{U}$ *is a compact subset of the space of continuous functions on* $\mathbb{D}$ *endowed with the topology of uniform convergence on compact subsets.*

Proof Let $f \in \mathcal{U}$. Since $f(0) = 0$ and $f'(0) = 1$, it follows from Schwarz's lemma that $f(\mathbb{D})$ cannot contain any disk with center 0 and radius greater than 1. Therefore there exist $z_1(f), z_2(f) \in \mathbb{C} \setminus f(\mathbb{D})$ such that $z_1(f)$ is in the circle of radius 2 and $z_2(f)$ is in the circle of radius 3. Now taking $z_3(f) = \infty$ we deduce from theorem 3.5.5 that $\mathcal{F}$ is a normal family. So the closure of the family is a compact set. Also, the limit of any convergent sequence of elements of the family is a holomorphic function that maps 0 to 0 and has derivative equal to 1 at 0. Thus this map is univalent, and therefore it belongs to the family. $\qquad\square$

Corollary 3.5.7 (Koebe's distortion lemma) *For each* $r < 1$, *there exists a constant* $C(r) \geq 1$ *such that if* $f \colon \mathbb{D} \to \mathbb{C}$ *is a univalent function then for every* z, w *in the disk of radius* r *centered at the origin we have*

$$\left| \log \frac{|f'(z)|}{|f'(w)|} \right| \leq C(r).$$

Proof Suppose, by contradiction, that there exists a sequence f_n of univalent functions on $\mathbb{D}$ and sequences z_n, w_n of points in the disk of radius r such that $|\log(|f'_n(z_n)|/|f'_n(w_n)|)| \to \infty$. For each n we can choose a function L_n that is the composition of a translation and a homothety such that the univalent function $g_n = L_n \circ f_n$ is normalized as in corollary 4.3.6. By the choice of L_n we have also

$$\left| \log \frac{|g_n'(z_n)|}{|g_n'(w_n)|} \right| \to \infty .$$

Passing to a subsequence if necessary, we may assume that $z_n \to z$, $w_n \to w$ and $g_n \to g$ uniformly on compact subsets. This is clearly a contradiction, because the derivative of g_n also converges to the derivative of g uniformly on compact subsets. $\qquad \square$

3.5.1 Fatou–Julia decomposition

Let us discuss now an important dynamical consequence of Montel's theorem that was obtained by Julia in [Ju]. Suppose that we are given a *rational map* $f : \widehat{\mathbb{C}} \to \widehat{\mathbb{C}}$, i.e. a map that can be written as the quotient of two polynomials, say $f(z) = P(z)/Q(z)$. Since f is a map of the Riemann sphere, its topological degree is the maximum of the algebraic degrees of P and Q. The *Fatou set* of f, denoted by $F(f)$, is the set of all points of the Riemann sphere having a neighborhood restricted to which the iterates of f form a normal family. The *Julia set* of f is the complement $J(f) = \widehat{\mathbb{C}} \setminus F(f)$ of the Fatou set.

Theorem 3.5.8 *If f is a rational function of degree $d \geq 2$ then*

(i) *the set of periodic points of f is dense in its Julia set $J(f)$;*

(ii) *the backward orbit of any point in $J(f)$ is dense in $J(f)$.*

Proof Let $p \in J(f)$ and let V be a neighborhood of p. We want to prove the existence of a periodic point of f in V. Let us suppose first that p is not a critical value of f, and also that it is not a fixed point of f. Then p has at least two distinct pre-images, p_1 and p_2. Also, since p is not a critical value there exist two local inverses of f in a small neighborhood $U \subset V$ of p, say $g_i : U \to U_i$, $i = 1, 2$, so that $f \circ g_i$ is the identity of U, $g_i(p) = p_i$ and $U_1 \cap U_2 = \emptyset$. For each $z \in U$ let M_z be the Möbius transformation that maps z to 0, $g_1(z)$ to 1 and $g_2(z)$ to ∞. If there exists a point $z \in U$ such that $f^n(z) = g_i(z)$ for some $n \in \mathbb{N}$ and some $i = 0, 1, 2$, then z is a periodic point of f and we are done. If no such point exists then the family of holomorphic maps $g_n(z) = M_z \circ f^n(z)$ is a normal family, since $0, 1, \infty$ are not in the image of any of these maps. Therefore the sequence $f^n(z) = M_z^{-1} g_n(z), z \in U$, is also a normal family. This is a contradiction because p belongs to the Julia set of f. This proves that p is accumulated by periodic points of f. If p is either a critical value or a fixed point we take a point q in the backward orbit of p that is not a critical value or a fixed point. This point q is also in the Julia set because the Julia set is totally invariant. Hence there exists a sequence of periodic points of f converging to q. The forward orbit of these points will contain points converging to p. This proves (i).

To prove (ii), let p be a point in the Julia set and let p_1, p_2, p_3 be three distinct points in the backward orbit of p. Let q be a point in the Julia set and U a neighborhood of q. By Montel's theorem there exists n such that $f^n(U)$ contains one of the points p_1, p_2, p_3. Therefore there exists a point in the backward orbit of p in any neighborhood of q. $\square$

Fatou and Julia proved that in fact $J(f)$ is the closure of the set of *repelling* periodic points of f (see section 3.7 below).

3.6 Uniformization of domains in the Riemann sphere

We are now in a position to state and prove a version of the celebrated *uniformization theorem* due to Poincaré, Klein and Koebe. Although the version below deals with domains in the Riemann sphere only, it is the most suitable for dynamical applications and also contains the *Riemann mapping theorem* as a special case. The statement (but not the proof) of the general uniformization theorem for abstract Riemann surfaces will be given in the Appendix.

Theorem 3.6.1 (Uniformization) *Let $U \subset \widehat{\mathbb{C}}$ be a domain whose complement contains at least three points. Then there exists a holomorphic covering map $\pi \colon \mathbb{D} \to U$. In particular, if U is simply connected then π is a diffeomorphism.*

Proof We may suppose that $U \subseteq \mathbb{C} \setminus \{0, 1\}$. Let $\lambda \colon \mathbb{D} \to \mathbb{C} \setminus \{0, 1\}$ be the covering map of theorem 3.5.1. By composing λ with a Möbius transformation we may assume that $z_0 = \lambda(0) \in U$. Let $\phi \colon \hat{U} \to U$ be the topological universal covering of U and choose a point $\hat{z}_0 \in \hat{U}$ with $\phi(\hat{z}_0) = z_0$. Let us consider in $\hat{U}$ the unique Riemann surface structure that makes ϕ holomorphic. Let us consider the family $\mathcal{F}$ of all holomorphic maps $F \colon \hat{U} \to \mathbb{D}$ with the following properties.

(i) $F(\hat{z}_0) = 0$.

(ii) F is a holomorphic covering map onto a domain $V_F \subset \mathbb{D}$.

(iii) There exists a holomorphic covering map $\Psi_F \colon V_F \to U$ such that $\Psi_F \circ F = \phi$.

Since $\hat{U}$ is simply connected, there exists a holomorphic map $\hat{\phi} \colon \hat{U} \to \mathbb{D}$ such that $\hat{\phi}(\hat{z}_0) = 0$ and $\lambda \circ \hat{\phi} = \phi$. Clearly $\hat{\phi}$ belongs to the family $\mathcal{F}$. Therefore $\mathcal{F} \neq \emptyset$.

Now we identify a neighborhood of $\hat{z}_0$ with a neighborhood of z_0 via ϕ and use this identification to measure the size of derivatives. Let $G_n \in \mathcal{F}$ be a sequence of maps such that $|G'_n(\hat{z}_0)|$ converges to $\sup\{|F'(\hat{z}_0)| : F \in \mathcal{F}\}$. Since this is a sequence of uniformly bounded holomorphic maps, we can assume, passing to a subsequence if necessary, that G_n converges uniformly on compact subsets to a holomorphic function G that maps z_0 to 0 and for which $|G'(z_0)| = \sup\{|F'(\hat{z}_0)| : F \in \mathcal{F}\}$. In particular G is non-constant. Hence the image of G is an open subset V_G of $\mathbb{D}$. Any $q \in V_G$ belongs to V_{G_n} if n is large enough, and it is easy to see that $\Psi_{G_n}(q)$ converges to some $\Psi_G(q)$; the map Ψ_G is holomorphic and satisfies $\Psi_G \circ G = \phi$. To deduce that G belongs to the family, it remains to prove that both G and Ψ_G are covering maps. We first show that Ψ_G is onto. Given $w \in U$, let $\hat{w} \in \hat{U}$ be such that $\phi(\hat{w})) = w$. Since $\Psi_G \circ G(\hat{w}) = \phi(\hat{w}) = w$, Ψ_G is onto. Next, let $B \subset U$ be a simply connected domain whose closure is compact and simply connected. Since ϕ is a covering map, $\phi^{-1}(B) = \cup B_i$, where the B_i are pairwise disjoint simply connected domains and the restriction of ϕ to each B_i is a diffeomorphism onto B. Since the closure of B_i is compact, the restriction of G_n to B_i converges uniformly to the restriction of G to B_i, which is a diffeomorphism onto a simply connected domain $B_i^G \subset V_G$. Since $\Psi_G \circ G = \phi$, it follows that the restriction of Ψ to B_i^G is a diffeomorphism onto B. Because G_n and Ψ_{G_n} are covering maps and $\Psi_{G_n} \circ G_n = \phi$, we have that either $B_i^{G_n} \cap B_j^{G_n} = \emptyset$ or $B_i^{G_n} = B_j^{G_n}$. So the same holds for B_i^G and B_j^G. Since again $\Psi_G \circ G = \phi$, we have that $\Psi_G^{-1}(B)$ is the union of the B_i^G, and the restriction of Ψ_G to each B_i^G is a diffeomorphism onto B. Hence Ψ_G is a covering map. However, $G^{-1}(B_i^G)$ is the union of all B_k such that $B_k^G = B_i^G$ and the restriction of G to each is a diffeomorphism onto B_i^G. The set of all B_i^G for all i and all $B \subset U$ is a covering of V_G and therefore G is a covering map. Hence $G \in \mathcal{F}$.

Next we will prove that $V_G = \mathbb{D}$, which will imply that G is a diffeomorphism and complete the proof of our theorem. Suppose, by contradiction, that there there exists $v \in \mathbb{D} \setminus V_G$. Let $f \colon \mathbb{D} \to \mathbb{D}$ be the holomorphic map constructed in proposition 3.4.1 that is a degree-2 covering map $\mathbb{D}\setminus\{c\} \to \mathbb{D}\setminus\{v\}$ and maps 0 to 0. Let W be the connected component of $f^{-1}(V_G)$ that contains 0. Then $f\colon W \to V_G$ is a covering map, and we can lift G to a holomorphic map $\hat{G} \to \hat{U} \to W$. Clearly $\hat{G}$ belongs to the family $\mathcal{F}$ and, since $f \circ \hat{G} = G$ and $|f'(0)| < 1$, we have that $|\hat{G}'(\hat{z}_0)| > |G'(\hat{z}_0)| = \sup\{|F'(\hat{z}_0)| : F \in \mathcal{F}\}$, which is the desired contradiction. $\qquad\square$

Thus we have proved that any domain of the Riemann sphere whose complement has at least three points has a hyperbolic metric. The Riemann sphere itself has a complete metric of constant curvature 1, the spherical metric. The complex plane has a metric of constant curvature 0, the Euclidean metric. The complement of two points in the sphere is covered by the plane and therefore has a complete metric of curvature 0 also.

3.6.1 Ring domains

An important example of uniformization is that of ring domains. A subdomain A of the Riemann sphere is called a *ring domain* if its complement has exactly two connected components. If each connected component is reduced to a single point then, as we have seen above, A has a complete metric of curvature zero. If one component is not a single point then there exists a holomorphic covering map $\pi\colon \mathbb{D} \to A$. Since the group of automorphisms of this covering is isomorphic to the fundamental group of A, it is generated by a unique Möbius transformation ϕ without fixed points in $\mathbb{D}$. So the fixed points of ϕ are in the boundary of $\mathbb{D}$. If ϕ has a unique fixed point p, we can conjugate ϕ by a Möbius transformation that maps $\mathbb{D}$ onto $\mathbb{H}$ and p into ∞ with the hyperbolic isometry $z \mapsto z + 1$ of $\mathbb{H}$. Hence a covering map $\mathbb{H} \to A$ exists whose automorphism group is $\{z \mapsto z + n : n \in \mathbb{Z}\}$. But this is precisely the automorphism group of the covering map of $\mathbb{D} \setminus \{0\}$. Thus A is conformally equivalent to $\mathbb{D} \setminus \{0\}$. Suppose now that the generator ϕ of the automorphism group of the covering $\pi\colon \mathbb{D} \to A$ has two fixed points in the boundary of $\mathbb{D}$. We can take a Möbius transformation mapping $\mathbb{D}$ onto $\mathbb{H}$ and the fixed points of ϕ into 0 and ∞. This Möbius transformation conjugates ϕ with the Möbius transformation $z \mapsto \lambda z$, where λ is a real number, say, greater than 1. Thus, composing π with the inverse of this Möbius transformation, we obtain a covering map $\pi\colon \mathbb{H} \to A$ whose automorphism group is $\{z \mapsto \lambda^n z : n \in \mathbb{Z}\}$, which is the automorphism group of the round annulus A_R of example 3.2.3, where

$\log R = 2\pi/\log \lambda$. Hence A is conformally equivalent to the annulus A_R. The image of the vertical geodesic connecting the fixed points of the automorphisms is a closed geodesic in A whose length is $\log \lambda$. It is easy to see that it minimizes the length of the closed curves in its homotopy class. Therefore the length of this geodesic is a conformal invariant of the ring. It is usual to consider a related invariant, called the *modulus* of the ring A, denoted mod A, which is the ratio of 2π and the length of the closed geodesic. Thus the modulus of A_R is $\log R$.

Proposition 3.6.2 *Let $f: A_1 \to A_2$ be a holomorphic covering map of degree d between two annuli. Then*

$$\mathrm{mod}\ A_1 = \frac{1}{d}\ \mathrm{mod}\ A_2\ .$$

Proof The map f is a local isometry between the hyperbolic metrics. Hence its restriction to the closed geodesic of A_1 is a degree-d covering of the closed geodesic of A_2. Thus the length of the closed geodesic of A_1 is d times the length of the closed geodesic of A_2. □

3.7 Dynamical applications

Let us move to some applications of these ideas to the dynamics of rational maps. The reader will find more systematic treatments of this theory in [B2], [CG], [Mi1] or [MNTU].

3.7.1 Periodic and critical points

Let $p \in \widehat{\mathbb{C}}$ be a periodic point of period n of a rational map $f: \widehat{\mathbb{C}} \to \widehat{\mathbb{C}}$. We say that p is *attracting* if $0 < |Df^n(p)| < 1$ and *super-attracting* if $Df^n(p) = 0$. We say that p is *repelling* if $|Df^n(p)| > 1$ and that it is an *indifferent* periodic point if $|Df^n(p)| = 1$. Although it will not be our object of study here, the *local* dynamics of a map near an attracting, super-attracting or repelling periodic point is fairly easy to analyze (see exercises 3.7, 3.11). This is not the case for indifferent periodic points, especially *irrationally* indifferent ones, i.e. those for which $Df^n(p)$ is not a root of unity. In particular, these are not always linearizable (see exercise 3.18 for an example).

The *basin* of an attracting (or super-attracting) periodic point is the set of points whose orbit is asymptotic to the orbit of the periodic point, i.e.

$$B(p) = \left\{ z \in \widehat{\mathbb{C}} : \lim_{k \to \infty} f^{kn+i}(z) = f^i(p) \right\}.$$

Clearly the basin of an attracting periodic point is an open subset of the Fatou set that contains the periodic orbit. The union of the n connected components of the basin containing the periodic orbit is called the *immediate basin*. The following basic fact was discovered by Fatou.

Theorem 3.7.1 *Let $f \colon \widehat{\mathbb{C}} \to \widehat{\mathbb{C}}$ be a rational map of degree $d \geq 2$. If p is an attracting fixed point of f then the immediate basin of p contains a critical point of f.*

Proof Since the basin is forward invariant, the complement must be backward invariant. Hence it must be infinite. Therefore the basin has a hyperbolic metric. If f had no critical point in the basin then the restriction of f to this basin would be a covering map; hence it would be a local isometry of the hyperbolic metric, but this is not possible because the fixed point is attracting. $\qquad\square$

Corollary 3.7.2 *If $f \colon \widehat{\mathbb{C}} \to \widehat{\mathbb{C}}$ is a rational function of degree $d \geq 2$ then the immediate basin of an attracting periodic orbit contains a critical point of f. In particular, the number of attracting periodic orbits of f is at most equal to $2d - 2$. If f is a polynomial then the number of attracting periodic orbits of f is at most equal to $d - 1$.*

Fatou proved that, given any rational map f, we can find arbitrarily small perturbations of f that turn at least half the indifferent periodic points of f into periodic attractors of the perturbed map. Combining this argument with the above corollary we deduce that the number of non-repelling periodic orbits of a rational map $f \colon \widehat{\mathbb{C}} \to \widehat{\mathbb{C}}$ of degree $d \geq 2$ is bounded by $4d - 4$. This bound was improved by M. Shishikura to $2d - 2$ using *quasiconformal maps*, which we will study in Chapter 4. Each repelling periodic point clearly belongs to the Julia set, and since there are only finitely many non-repelling periodic points we deduce that the set of repelling periodic points is a dense subset of the Julia set. This yields the Fatou–Julia decomposition to which we alluded in subsection 3.5.1. Furthermore, given any neighborhood W of a periodic repelling orbit, there is a neighborhood $V \subset W$ of this orbit such that $f(V) \supset V$. Thus $f^{k+1}(V) \supset f^k(V)$. As we have seen, $\cup_{n \in \mathbb{N}} f^n(V)$ contains the whole sphere except possibly for two points, which are the *exceptional points* of f (see exercise 3.6). In particular this union contains the Julia set which is compact. Hence, since this family of open sets is increasing, there exists k such that $f^k(V)$ contains the Julia set. Using Montel's theorem once again we deduce that, given any open set U that intersects the Julia set, the union of a finite number of iterates of U covers the Julia set. This is so because U must contain a repelling periodic point and a finite union of iterates of U is a neighborhood of the corresponding periodic orbit.

The *post-critical* set of a rational map f is the closure of the forward orbits of the critical values of f, in other words

$$P(f) \;=\; \overline{\bigcup_{n=1}^{\infty} f^n(C(f))} \;,$$

where $C(f)$ is the set of critical points of f. This is clearly a compact forward-invariant set. In general the *critical* set of an iterate of a map f is much larger than the critical set of f. By contrast, the post-critical set of any iterate of f is equal to the post-critical set of f. If the post-critical set has only two points then, by conjugating the map with a Möbius transformation, we can assume that these points are $\{0, \infty\}$ and we see that the corresponding map must be $z \mapsto z^d$. If the post-critical set has three or more points, we can consider the hyperbolic metric on the connected components of the complement of $P(f)$ and, by the Schwarz lemma, we have that in each component either f strictly contracts the hyperbolic metric or it is a local isometry. For points in the Julia set we have the following.

Theorem 3.7.3 *Let f be a rational map whose post-critical set has more than two points. If z is a point in the Julia whose forward orbit does not intersects the post-critical set then the norm of the derivatives $Df^n(z)$ in the hyperbolic metric of the complement of $P(f)$ tends to ∞ as $n \to \infty$.*

Proof Let $Q_n = f^{-n}(P(f))$. Since $f(P(f)) \supset P(f)$, we have that $Q_{n+1} \supset Q_n \supset P(f)$. By Montel's theorem, any neighborhood of z intersects Q_n for some n. Hence the spherical distance between z and Q_n converges to zero as $n \to \infty$. Therefore the norm of the derivative at z of the inclusion $i_n \colon \widehat{\mathbb{C}} \setminus Q_n \to \widehat{\mathbb{C}} \setminus P(f)$, with respect to the hyperbolic metric, tends to zero as $n \to \infty$. However, $f^n \colon \widehat{\mathbb{C}} \setminus Q_n \to \widehat{\mathbb{C}} \setminus P(f)$ is a local isometry of the hyperbolic metrics. This proves the theorem. $\qquad\square$

Corollary 3.7.4 *Every periodic point in the Julia set that does not belong to the post-critical set is repelling.*

Corollary 3.7.5 (Hyperbolicity) *If each critical point of f is either periodic or belongs to the basin of a periodic attractor then there exists n such that the norm of $Df^n(z)$ in the spherical metric is greater than 1 for all z in the Julia set.*

Corollary 3.7.6 *If f is a rational map such that the post-critical set is finite and all periodic points in the post-critical set are repelling then the Julia set is the whole sphere.*

The above theorem combined with the Koebe's distortion lemma has the following important consequence whose proof can be found in [McM1, p. 42].

Theorem 3.7.7 *If f is a rational map of degree greater than or equal to 1 then either*

(i) *the Julia set of f is the whole sphere and f is ergodic with respect to the Lebesgue measure, i.e. each totally invariant set has either full Lebesgue measure or zero Lebesgue measure, or*

(ii) *the spherical distance from the iterates of x to the post-critical set of f tends to zero for Lebesgue-almost points in the Julia set.*

Corollary 3.7.8 *If each critical point of a rational map f is periodic, is in the basin of a periodic attractor or is in the backward orbit of a repelling periodic point then either the Julia set is the whole sphere and f is ergodic or the Julia set has zero Lebesgue measure.*

3.7.2 The components of the Fatou set

Let $f\colon \widehat{\mathbb{C}} \to \widehat{\mathbb{C}}$ be a rational map of degree $d \geq 2$. Since the Fatou set of f is totally invariant, f maps each component (i.e. each connected component) of the Fatou set onto a component. Hence the forward orbit of a component is a sequence of components. We say that the component is periodic of period n if its forward orbit has exactly n components and is therefore fixed under f^n. Using Schwarz's lemma, we will describe all possible dynamical behaviors in the periodic components. Let U be a periodic component of a rational map f. By passing to an iterate we may assume that it is a fixed component. Clearly the complement of U has infinitely many points, since it contains the Julia set. Therefore we may consider the hyperbolic metric on U and, by Schwarz's lemma, either $f|U$ is a local isometry or it is a strict contraction of the hyperbolic metric.

Lemma 3.7.9 *If there exists $z_0 \in U$ such that the sequence $\{f^n(z_0)\}$ converges to the boundary of U then it converges to a fixed point w of f and $f^n(z) \to w$ for all $z \in U$.*

Proof Let α be a smooth curve connecting z_0 to $f(z_0)$ and α_n be the curve $f^n \circ \alpha$. By Schwarz's lemma, the hyperbolic length of α_n is a non-increasing sequence. Hence the hyperbolic distance between $f^n(z_0)$ and its image $f(f^n(z_0))$ is uniformly bounded and, since the first sequence converges to the boundary of U, their Euclidean distance must tend to zero. Hence if $w = \lim_{i \to \infty} f^{n_i}(z_0)$ then w is a fixed point of f. Since f has a finite number of fixed points, the whole sequence must converge to a unique fixed point because the hyperbolic distance between two points in small neighborhoods of different fixed points is very large. Also, $f^n(z)$ must converge to this fixed point since the Euclidean distance between $f^n(z_0)$ and $f^n(z)$ converges to zero because their hyperbolic distance is uniformly bounded. □

Lemma 3.7.10 *If $f: U \to U$ is a strict contraction of the hyperbolic metric and there exist $w \in U$ and a sequence of iterates $f^{n_i}(z_0) \to w$, then w is a fixed point and $f^n(z) \to w$ for all $z \in U$.*

Proof Let α be a smooth curve connecting z_0 to $f(z_0)$. Let B be the hyperbolic ball centered on w with radius twice the length of the curve α. Since the closure of B is compact, the norm of the derivative of f with respect to the hyperbolic metric is smaller than a number $\lambda < 1$ in B and also the hyperbolic metric is commensurable with the Euclidean metric in B. Whenever $f^{n_i}(z_0)$ is very close to w, the curve $\alpha_{n_i} = f^{n_i}(\alpha)$ will be completely inside B and, therefore, the hyperbolic length of its image will decrease by a factor λ. This implies that the hyperbolic lengths of the full sequence α_n converge to zero. Hence the Euclidean distance between $f^{n_i}(z_0)$ and its image converges to zero, which implies again that w is a fixed point. This fixed point attracts the iterates of all other points, again by contraction of the hyperbolic metric. □

Lemma 3.7.11 *Let* Γ *be a discrete group of hyperbolic isometries of* $\mathbb{D}$. *If* Γ *is not commutative then the set of Möbius transformations* $\phi\colon \mathbb{D} \to \mathbb{D}$ *such that* $\phi\Gamma\phi^{-1} \subset \Gamma$ *is discrete.*

Proof Suppose, by contradiction, that there exists a sequence ϕ_n of such Möbius transformations converging to ϕ. Then $\psi_n = \phi^{-1} \circ \phi_n$ is a sequence of distinct Möbius transformations converging to the identity and satisfying $\psi_n \circ \Gamma \circ \psi_n^{-1} \subset \Gamma$. Hence, for each $\alpha \in \Gamma$, $\psi_n \circ \alpha \circ \psi^{-1} = \alpha$ for all sufficiently large n. Since two Möbius transformations commute if and only if they have the same fixed points, we have that α and h_n have the same fixed points for large enough n. Hence any two Möbius transformations in Γ have the same fixed points, and the group Γ is commutative. $\square$

Lemma 3.7.12 *Suppose that the automorphism group of a covering* $\pi\colon \mathbb{D} \to U$ *is non-commutative. If* $f\colon U \to U$ *is a covering map and there exist* $z, w \in U$ *such that* $f^{n_i}(z) \to w$ *then* f *is an automorphism of finite order, i.e. some iterate of* f *is the identity.*

Proof Let $\hat{z}$ and $\hat{w}$ in $\mathbb{D}$ project down to z and w. Let ϕ_i be a lift of f^{n_i} that maps $\hat{z}$ near $\hat{w}$. Then ϕ_i belongs to a compact set of Möbius transformations of $\mathbb{D}$. Since each ϕ_i is the lift of an endomorphism of U we have that $\phi_i\Gamma\phi_i^{-1} \subset \Gamma$, where Γ is the automorphism group of the covering. By lemma 3.7.11, this sequence is finite. Hence there exist j, i, with $n_j > n_i$, such that $\phi_j = \phi_i$. But this implies that $f^{n_j} = f^{n_i}$ with $n_j > n_i$. If the degree of f as a covering were greater than 1 then the degree of f^{n_j} would be greater than the degree of f^{n_i}. Hence f is an automorphism of U of finite order. $\square$

Theorem 3.7.13 *Let* U *be a fixed Fatou component of a rational map* f *of degree* $d \geq 2$. *Then the dynamics of* f *on* U *is described by one of the possibilities below (see figure 3.5).*

   (i) Attracting basin. *There is an attracting fixed point* w *in* U *and* U *is the immediate basin of* w.

   (ii) Super-attracting basin. *There is a super-attracting fixed point* w *in* U *and* U *is its immediate basin.*

   (iii) Parabolic basin. *There is a fixed point* w *in the boundary of* U *and* $f^n(z) \to w$ *for all* $z \in U$.

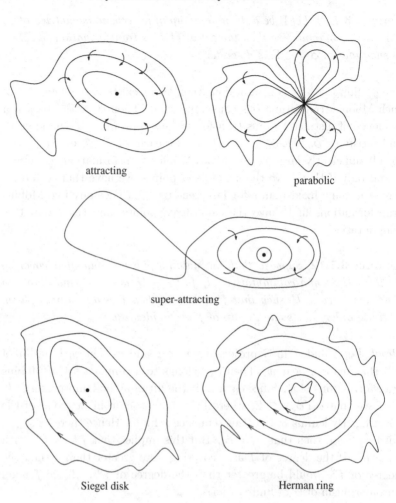

Fig. 3.5. Stable Fatou components.

   (iv) Siegel disk. *There is a holomorphic diffeomorphism $h\colon U \to \mathbb{D}$ that conjugates f with an irrational rotation.*

   (v) Herman ring. *There is a holomorphic diffeomorphism $h\colon U \to A_R$ that conjugates f with an irrational rotation of the annulus.*

Proof If there is a positive orbit in U converging to the boundary of U we arrive in the third case, by lemma 3.7.9. Otherwise, $f^{n_i}(z) \to w \in U$ for some $z, w \in U$. If furthermore f contracts the hyperbolic metric we are now in the first two cases, by lemma 3.7.10. The final possibility is the existence of z, w as above and that f is a hyperbolic local isometry and a covering map. By lemma 3.7.12 the group of automorphisms of the covering $\pi \colon \mathbb{D} \to U$ must be commutative since f cannot be of finite order. If the automorphism group of the covering is trivial, then π is a diffeomorphism for which $\pi^{-1} \circ f \circ \pi$ is a hyperbolic isometry having an orbit with non-empty ω-limit set. Hence it is holomorphically equivalent to a rotation. This rotation must be irrational because f cannot have finite order and we are in the fourth case. Now U cannot be conformally equivalent to $\mathbb{D} \setminus \{0\}$, otherwise the bounded component of its complement would be an isolated point of the Julia set. Hence the automorphism group of π must be generated by a hyperbolic isometry with two fixed points in the boundary of $\mathbb{D}$. Thus U is a cylinder and f must be an automorphism of U. Since the group of automorphisms of the annulus A_R is the group of rotations and f cannot be a finite-order automorphism, f must be holomorphically conjugate to a rotation of A_R. $\qquad\square$

Fatou proved that if f is a fixed point of a holomorphic map and p is in the boundary of an open set contained in the basin of p then either $|f'(p)| < 1$ or $f'(p) = 1$. Douady and Sullivan gave a simple geometric proof of this statement using Schwarz's lemma, see [Mi1].

Notice that both Siegel disks and Herman rings are foliated by circles and that this foliation is dynamically defined: each circle is the closure of an orbit of the iterate that fixes the component. Hence the foliation is preserved by topological conjugacies. The basin of a super-attracting periodic point also has a dynamically defined foliation by circles in a neighborhood of the critical periodic point. If f is the iterate that fixes the component and x is a point in the component of the critical fixed point then the leaf through x is the closure of the set $\{y \in U; f^n(y) = f^n(x), n \in \mathbb{N}\}$. If x is close to the fixed critical point of f then this leaf is a circle, since f restricted to a neighborhood of the critical fixed point is holomorphically conjugate to $z \mapsto z^k$ where $k - 1$ is the multiplicity of the critical fixed point. This is the so-called Böttcher theorem, whose proof can be found in [Mi1]. In the Böttcher coordinates each leaf is a round circle. We can decompose the full basin U into dynamically defined curves, taking the pre-images of these circles by iterates of f. Of course, if such a circle contains a critical value of an iterate, its pre-image under that iterate will no longer be a circle or a union of circles but will have a singularity at the corresponding critical point.

3.7.3 Siegel disks, Herman rings and the post-critical set

Let us now illustrate the use of normal families in the study of some basic aspects of the dynamical behavior of a rational map at the boundary of a Siegel disk or Herman ring. First, we give a general result concerning the inverse branches of such maps.

Lemma 3.7.14 *Let* $f : \widehat{\mathbb{C}} \to \widehat{\mathbb{C}}$ *be a rational map of degree* ≥ 2, *and let* $V \subseteq \widehat{\mathbb{C}}$ *be a domain. Fix a sequence* $(n_k)_{k \geq 0}$ *of non-negative integers, and for each* $k \geq 0$ *let* $g_k : V \to \widehat{\mathbb{C}}$ *be a univalent inverse branch of* f^{n_k}. *Then* $\{g_k\}$ *is a normal family in* V.

Proof First we find two periodic points ζ and w of f with minimal periods p and q respectively, where $p, q \geq 3$ are relatively prime. Then the periodic cycles $O(\zeta) = \{\zeta, f(\zeta), \ldots, f^{p-1}(\zeta)\}$ and $O(w) = \{w, f(w), \ldots, f^{q-1}(w)\}$ are disjoint. Note that no point in $V \setminus O(\zeta)$ can be mapped into $O(\zeta)$ by any g_k: if this were to happen for some $z \in V \setminus O(\zeta)$, we would have $g_k(z) = f^i(\zeta)$ for some $k \geq 0$ and some $0 \leq i \leq p - 1$, and this would imply that

$$z = f^{n_k} \circ g_k(z) = f^{n_k - i}(\zeta) \in O(\zeta),$$

a contradiction. This shows that $\{g_k\}$ is a normal family in $V \setminus O(\zeta)$. Similarly, $\{g_k\}$ is a normal family in $V \setminus O(w)$. Therefore, $\{g_k\}$ is normal in $V \setminus O(\zeta) \cup V \setminus O(w) = V$. □

It is worth observing that, whenever V in the above lemma is simply connected and n is such that f^n has no critical points in V, every connected component of $f^{-n}(V)$ is mapped bijectively onto V. Thus each inverse branch of f^n in V is a univalent map onto a component of $f^{-n}(V)$, and such components are simply connected as well.

Theorem 3.7.15 *Let* $U \subseteq F(f)$ *be a Siegel disk or a Herman ring of a rational map* $f : \widehat{\mathbb{C}} \to \widehat{\mathbb{C}}$. *Then* ∂U *is contained in the closure of the post-critical set,* $P(f)$, *of* f.

Proof We will argue by contradiction. Suppose there exist $\zeta \in \partial U$ and a disk $D = D(\zeta, r)$ such that $D \cap P(f) = \emptyset$. We assume that r is sufficiently small that D does not contain the whole Julia set $J(f)$. Recall that $f|_U : U \to U$ is conjugate to an irrational rotation. Hence so is its inverse $g : U \to U$. For each $n \geq 1$, let $g_n : D \to \widehat{\mathbb{C}}$ be the univalent inverse branch of f^n that agrees with g^n on $D \cap U$. By lemma 3.7.14,

$\{g_n\}$ is a normal family in D. Choose a subsequence $n_k \to \infty$ such that $g^{n_k} \to \mathrm{id}_U$ uniformly on compact subsets of U (this is possible because g is conjugate to an irrational rotation). Then, by Vitali's theorem (see exercise 2.7), $\{g_{n_k}\}$ converges uniformly on compact subsets of D to the identity; in particular, $g_{n_k}|_{D'} \to \mathrm{id}_{D'}$ uniformly, where $D' = D(\zeta, r/2) \subset D$. This implies that

$$g_{n_k}(D') \supseteq D'' = D(\zeta, r/4)$$

for every sufficiently large k. Therefore

$$f^{n_k}(D'') \subset f^{n_k} \circ g_{n_k}(D') = D'.$$

But we know that, if n is sufficiently large, $f^n(D'')$ must contain $J(f)$. Hence $J(f) \subset D'$, a contradiction. $\qquad\square$

As an application of this result, let us prove the following slightly more general version of corollary 3.7.6. We shall use also Sullivan's no-wandering-domains theorem, to be proved in Chapter 4.

Theorem 3.7.16 *Let $f : \widehat{\mathbb{C}} \to \widehat{\mathbb{C}}$ be a rational map of degree ≥ 2, all of whose critical points are pre-periodic but not periodic. Then $J(f) = \widehat{\mathbb{C}}$.*

Proof The hypothesis tells us that $P(f)$ is finite. Let $U \subseteq F(f)$ be a stable component. Then U cannot be a super-attracting domain, for there are no periodic critical points. It cannot be an attracting domain, otherwise by theorem 3.7.1 some critical point would fall in the basin of its attracting cycle, and the forward orbit of such critical point would be infinite. For similar reasons, U cannot be a parabolic domain either. Finally, U cannot be a Siegel disk or a Herman ring, otherwise by theorem 3.7.15 the boundary of U would be accumulated by $P(f)$, which then would have to be infinite. Therefore there are no stable components in the Fatou set. But, by Sullivan's theorem, every component of the Fatou set is eventually mapped to a stable component. This shows that $F(f) = \emptyset$, and so $J(f) = \widehat{\mathbb{C}}$. $\qquad\square$

3.7.4 Examples

Here are some examples of rational maps and their Julia sets.

Example 3.7.17 *An interval.* Let $f : \widehat{\mathbb{C}} \to \widehat{\mathbb{C}}$ be the quadratic polynomial $f(z) = z^2 - 2$. The Julia set of this map is the closed interval $[-2, 2] \subset \widehat{\mathbb{C}}$. This is left as an exercise for the reader (see exercise 3.15).

Example 3.7.18 *A Jordan curve.* Let $f : \widehat{\mathbb{C}} \to \widehat{\mathbb{C}}$ be the polynomial $f(z) = z^d$, where $d \geq 2$. It is an easy exercise to check directly that $J(f)$ is the unit circle. Both components of $F(f) = \widehat{\mathbb{C}} \setminus J(f)$ are super-attracting domains, 0 and ∞ being super-attracting fixed points.

Example 3.7.19 *A Cantor set.* Let $f : \widehat{\mathbb{C}} \to \widehat{\mathbb{C}}$ be the quadratic polynomial $f(z) = z^2 + c$, where $|c|$ is large. As we have already seen in section 3.3, the Julia set in this case is a Cantor set (and the dynamics of f restricted to $J(f)$ is conjugated to a one-sided shift on two symbols). The map f restricted to a neighborhood of the Julia set is an example of a *Cantor repeller*; see section 3.9 below.

Example 3.7.20 *A dendrite.* Let $f : \widehat{\mathbb{C}} \to \widehat{\mathbb{C}}$ be the quadratic polynomial $f(z) = z^2 + i$. Here the only critical points are ∞, which is super-attracting, and 0, which is pre-periodic but not periodic:

$$0 \quad \mapsto \quad i \quad \mapsto \quad -1+i \quad \mapsto \quad -i \quad \mapsto \quad -1+i \ .$$

Hence f is critically finite. By theorem 3.7.15, f has no Siegel disks or Herman rings (in fact, polynomial maps never have Herman rings, see exercise 3.16). There are no parabolic components and no attracting components either, because $P(f)$ is finite. Therefore $\widehat{\mathbb{C}} \setminus J(f) = F(f)$ has only one component, namely the super-attracting domain containing ∞. Therefore the Julia set $J(f)$ is compact and simply connected. We claim that it is also connected. This can be seen as follows. If $R > 0$ is sufficiently large, the disk $D = D(0, R)$ is mapped *over* itself, i.e. $f(D) \supset \overline{D}$, as a degree-2 map with a critical point at 0. Equivalently, the disk $\widehat{\mathbb{C}} \setminus D$ is mapped strictly into itself as a degree 2 branched covering map that is branched at ∞. It is clear that $f^{-(n+1)}(D) \subset f^{-n}(D)$ for all $n \geq 0$. From these facts it is easy to deduce that

$$F(f) = \bigcup_{n \geq 0} f^{-n}(\widehat{\mathbb{C}} \setminus D) \ .$$

In other words, the Fatou set is an increasing union of simply connected open sets, and therefore it is simply connected as well. But this means that its complement $J(f)$ is connected as claimed. It follows that $J(f)$ is a *dendrite*, a compact, connected and simply connected subset of the complex plane having empty interior. This is clearly seen in figure 3.6.

Example 3.7.21 *The whole sphere.* Now let $f : \widehat{\mathbb{C}} \to \widehat{\mathbb{C}}$ be the rational map $f(z) = 1 - 2/z^2$. This rational map has critical points at

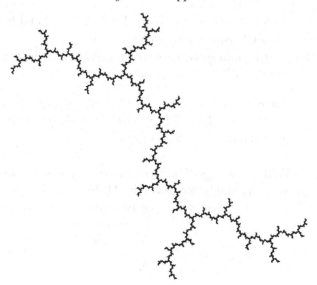

Fig. 3.6. Julia set for $z \mapsto z^2 + i$.

0 and ∞. We have

$$0 \; \mapsto \; \infty \; \mapsto \; 1 \; \mapsto \; -1 \,,$$

and the point -1 is a repelling fixed point (indeed, $f'(-1) = -4$). Thus f is critically finite, with all critical points pre-periodic but not periodic. By theorem 3.7.16, $J(f)$ is the whole Riemann sphere. Many other examples of this type exist. A very interesting family of rational maps with Julia sets equal to the whole sphere is given by the so-called *Lattès examples*, constructed with the help of the Weierstrass elliptic function (see [B2, ch. 4] for a detailed exposition).

Example 3.7.22 *A map with a Siegel disk.* We again consider a quadratic polynomial, this time of the form $f(z) = \lambda z + z^2$. Here we suppose that $\lambda = e^{2\pi i \theta}$, where θ is a *Diophantine* number, that is to say, an irrational number for which there exist constants $C, \beta > 0$ such that

$$\left| \theta - \frac{p}{q} \right| \geq \frac{C}{q^{2+\beta}} \tag{3.1}$$

for all rationals $p/q \in \mathbb{Q}$. By a difficult theorem due to C. Siegel [Sie] (see also [CG, pp. 43–6]), the map f is linearizable at $z = 0$; in other words, there exists an analytic map h defined on an open neighborhood V of 0 with the following properties.

(1) The map h is univalent and normalized: $h(0) = 0$ and $h'(0) = 1$.
(2) The open set V is invariant under f: $f(V) \subseteq V$.
(3) The map f is conjugated by h to a rotation: $h \circ f = R_\theta \circ h$ in V, where $R_\theta(z) = e^{2\pi i \theta}$ is a rotation by $2\pi\theta$.

From (3), we have of course that $f^n = h^{-1} \circ R_\theta^n \circ h$ in V for all n, so $\{f^n\}_{n \geq 0}$ is normal in V. Therefore $V \subseteq F(f)$, and the component of the Fatou set containing V is a Siegel disk.

Example 3.7.23 *A map with a Herman ring.* The first example of such a map, not surprisingly, was given by M. Herman. Let us consider for each $\alpha \in \mathbb{R}$ and each real $\beta > 1$ the Blaschke product $f_{\alpha,\beta} : \widehat{\mathbb{C}} \to \widehat{\mathbb{C}}$ given by

$$f_{\alpha,\beta} = e^{2\pi i \alpha} z^2 \left(\frac{z - \beta}{1 - \overline{\beta} z} \right).$$

Then $f_{\alpha,\beta}$ leaves the unit circle invariant, and $f_{\alpha,\beta}|_{S^1}$ is a real analytic circle diffeomorphism. The rotation number of this circle diffeomorphism varies continuously with both α and β. Moreover, when β is very large, $f_{\alpha,\beta}|_{S^1}$ is uniformly close to the rotation R_α, and we can choose $\alpha = \alpha(\beta)$ in such a way that the rotation number θ of $f_{\alpha,\beta}|_{S^1}$ takes whatever value we like. We choose it so that the rotation number is Diophantine, i.e. satisfies (3.1). So our circle diffeomorphism is a small perturbation of a rigid rotation, and its rotation number is Diophantine. By a theorem due to V. Arnold, under these conditions the analytic diffeomorphism $f_{\alpha,\beta}|_{S^1}$ is analytically conjugate to a rotation. In other words, there exist a doubly connected open neighborhood V of S^1 in the complex plane and an analytic map $h : V \to \mathbb{C}$ with the following properties.

(1) The map h is univalent and preserves the unit circle: $h(S^1) = S^1$.
(2) The open set V is invariant under f: $f(V) \subseteq V$.
(3) The map f is conjugated by h to a rotation: $h \circ f = R_\theta \circ h$ in V, where $R_\theta(z) = e^{2\pi i \theta}$ is a rotation by $2\pi\theta$.

This is an exact analogue of Siegel's theorem, and in fact both theorems can be seen as special cases of the famous *KAM theorem*, see for instance [KH]. Just as in the previous example we have $V \subseteq F(f_{\alpha,\beta})$, and the component W of $F(f)$ containing V is such that $f|_W$ is conjugate to the rotation R_θ. This component cannot be a Siegel disk because both 0 and ∞ are super-attracting fixed points, so it must be a Herman ring. A different, elegant, proof of the existence of rational maps having Herman

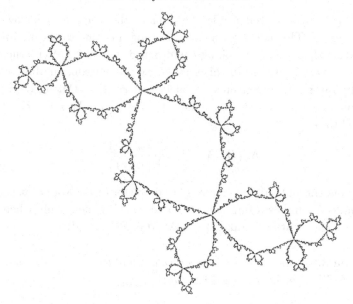

Fig. 3.7. Douady's rabbit: the Julia set for $z \mapsto z^2 + c$, with $c = -0.122\,561 + 0.744\,862\,i$.

rings was obtained by M. Shishikura through the use of *quasiconformal surgery*; see Chapter 4.

Example 3.7.24 *An infinitely connected Julia set.* Here is an example in which the Fatou set has infinitely many components. Again, we look at a quadratic polynomial of the form $f(z) = z^2 + c$. We choose $c \neq 0$ so that $f^3(0) = 0$, i.e. so that the (finite) critical point of f is periodic of period 3. In other words, c is a solution of the following equation:

$$c^3 + 2c^2 + c + 1 = 0 \; .$$

One such solution is $c \simeq -0.122\,561 + 0.744\,862\,i$. With this value of c, the origin is a super-attracting periodic point. Accordingly, there is a cycle of period 3 of super-attracting domains. Infinitely many components of the Fatou set are eventually mapped into this cycle. The Julia set is depicted in figure 3.7; it is known as *Douady's rabbit*.

3.8 Conformal distortion

As the Riemann mapping theorem shows, the geometric behavior of a conformal mapping near the boundary of its domain can be quite wild. In contrast, well inside the domain the mapping's behavior is rather tame. This is the geometric meaning, in some sense, of Koebe's distortion theorem. Controlling conformal distortion is often crucial in dynamical applications. In this section, we present some useful results.

Suppose that we are given a univalent map $\phi : V \to \mathbb{C}$ defined on some domain $V \subseteq \mathbb{C}$, and a closed set $D \subset V$. We define the *non-linearity of* ϕ *in* D to be

$$N_\phi(D) = |D| \sup_{z \in D} \left| \frac{\phi''(z)}{\phi'(z)} \right| .$$

Here, and throughout, $|D|$ denotes the diameter of D. Note that $N_\phi(D)$ is non-decreasing as a function of D and vanishes identically when ϕ is linear. This notation is taken from [McM5]; see also [dF1].

Lemma 3.8.1 *Let* $\phi : V \to \mathbb{C}$ *be univalent and let* $D \subseteq V$ *be closed and convex. Then for all* $z_1, z_2 \in D$ *we have*

$$\exp(-N_\phi(D)) \leq \left| \frac{\phi'(z_1)}{\phi'(z_2)} \right| \leq \exp(N_\phi(D)) . \tag{3.2}$$

Proof Let $[z_1, z_2] \subseteq D$ be the straight line segment joining z_1 to z_2. Then

$$\log \left| \frac{\phi'(z_1)}{\phi'(z_2)} \right| \leq \int_{[z_1, z_2]} \left| \frac{\phi''(z)}{\phi'(z)} \right| |dz|$$

$$\leq |z_1 - z_2| \sup_{z \in D} \left| \frac{\phi''(z)}{\phi'(z)} \right| \leq N_\phi(D) ,$$

from which (3.2) follows. $\qquad\qquad\qquad\qquad\qquad\qquad\qquad\qquad\square$

The non-linearity $N_\phi(D)$ can be bounded with the help of Koebe's pointwise estimate

$$\left| \frac{\phi''(z)}{\phi'(z)} \right| \leq \frac{4}{\text{dist}(z, \partial V)} .$$

This estimate follows easily from exercise 2.6. Hence we have

$$N_\phi(D) \leq \frac{4|D|}{\text{dist}(z, \partial V)} .$$

This inequality combined with lemma 3.8.1 yields immediately the following result.

Lemma 3.8.2 *Let $\phi : V \to \mathbb{C}$ and $D \subseteq V$ be as in lemma 3.8.1. Then for all $z_1, z_2 \in D$ we have*

$$\frac{1}{K} \leq \left| \frac{\phi'(z_1)}{\phi'(z_2)} \right| \leq K , \tag{3.3}$$

where $K = \exp\left(4|D|/\operatorname{dist}(D, \partial V)\right)$. $\square$

It turns out that we can even replace derivatives by difference quotients. More precisely, we have the following result.

Lemma 3.8.3 *Let $\phi : V \to \mathbb{C}$ and $D \subseteq V$ be as in lemma 3.8.1. Then for all $x, y, z \in D$ we have*

$$C^{-1}|\phi'(z)| \leq \frac{|\phi(x) - \phi(y)|}{|x - y|} \leq C|\phi'(z)| ,$$

where C depends only on the ratio $|D|/\operatorname{dist}(D, \partial V)$.

Proof To get the upper estimate, we note that

$$
\begin{aligned}
|\phi(x) - \phi(y)| &\leq \int_{[x,y]} |\phi'(\zeta)| \, |d\zeta| \\
&\leq |x - y| \sup_{\zeta \in [x,y]} |\phi'(\zeta)| \\
&\leq K|\phi'(z)| \, |x - y| ,
\end{aligned}
$$

by the inequality in lemma 3.8.2. To get the lower estimate, we use Koebe's one-quarter theorem. Let

$$r = \min\{\tfrac{1}{2}|x - y| , \operatorname{dist}(D, \partial V)\} . \tag{3.4}$$

Then ϕ is univalent in the disk $D(x, r)$ of radius r about x, and Koebe's theorem tells us that

$$\phi(D(x, r)) \supseteq D(\phi(x), \tfrac{1}{4}r|\phi'(x)|) .$$

Since y lies outside $D(x, r)$ and ϕ is injective, we have $\phi(y) \notin \phi(D(x, r))$, and therefore

$$|\phi(x) - \phi(y)| \geq \tfrac{1}{4}r|\phi'(x)| . \tag{3.5}$$

However, since $|x - y| \leq |D|$ we see from (3.4) that

$$\frac{r}{|x - y|} \geq \min\left\{\frac{1}{2}, \frac{\operatorname{dist}(D, \partial V)}{|D|}\right\} = C_1 .$$

Inserting this information into (3.5) we get

$$\frac{|\phi(x) - \phi(y)|}{|x - y|} \geq \frac{C_1}{4}|\phi'(x)| \geq \frac{C_1}{4K}|\phi'(z)| \, ,$$

where once again we have used the inequality in lemma 3.8.2. Comparing constants in the upper and lower estimates, we deduce the inequalities in the statement provided that $C = \max\{K, 4K/C_1\}$, and this is clearly a function of $|D|/\operatorname{dist}(D, \partial V)$ only. $\qquad\square$

3.9 Thermodynamics of Cantor repellers

In this final section, we shall present an application of the conformal distortion techniques seen in this chapter to the study of a special class of *conformal repellers*. This will provide us with an opportunity to introduce the reader to two important tools coming from ergodic theory: the concepts of *pressure* and *Gibbs* or *equilibrium states*. These in turn have been inspired by statistical mechanics. Our treatment here is strongly influenced by the elegant exposition in [Zi]. For further information on this subject, the reader should consult [BKS], as well as the more general references [KH], [Wa].

3.9.1 Cantor repellers

By a *conformal repeller*, we mean a triple (U, f, V) where $U, V \subseteq \mathbb{C}$ are open sets with U compactly contained in V and $f : U \to V$ is a holomorphic covering map with the following properties.

(1) There exist $C > 0$ and $\lambda > 1$ such that $|(f^n)'(z)| \geq C\lambda^n$ for all $z \in J_f = \cap_{n \geq 0} f^{-n}(V)$.

(2) The map f is locally eventually onto J_f: for all open sets $\mathcal{O}$ with $\mathcal{O} \cap J_f \neq \emptyset$ we have $f^n(\mathcal{O} \cap J_f) \supseteq J_f$.

The set J_f is called the *limit set* of the repeller. The reader can check as an exercise that J_f is a compact totally invariant set.

Various examples of conformal repellers arise very naturally in complex dynamics. One of the simplest is provided by the quadratic map $f(z) = z^2$, with V the annulus $\{z : \frac{1}{2} < |z| < 2\}$, say, and $U = f^{-1}(V)$. In this case, the limit set J_f is the unit circle. Another simple example is the Markov map we saw in subsection 3.3.1. What we want to study here is a strong generalization of this second example.

Definition 3.9.1 *A Cantor repeller (see figure 3.8) consists of two open sets $U, V \subseteq \mathbb{C}$ and a holomorphic map $f : U \to V$ satisfying the following conditions:*

(i) *the domain U is the union of Jordan domains $U_1, U_2, \ldots, U_m$ (for some $m \geq 2$) having pairwise disjoint closures;*

(ii) *the codomain V is the union of Jordan domains $V_1, V_2, \ldots, V_M$ (for some $M \geq 1$) having pairwise disjoint closures;*

(iii) *for each $i \in \{1, 2, \ldots, m\}$ there exists $j(i) \in \{1, 2, \ldots, M\}$ such that $f|_{U_i}$ maps U_i conformally onto $V_{j(i)}$;*

(iv) *we have $\overline{U} \subset V$;*

(v) *the limit set $J_f = \cap_{n \geq 0} f^{-n}(V)$ has the locally eventually onto property, as previously stated.*

Remark 3.9.2 *Note that the limit set J_f of a Cantor repeller is a compact, perfect and totally disconnected set, i.e. a Cantor set, hence the name "Cantor".*

Proposition 3.9.3 *Every Cantor repeller is a conformal repeller.*

Proof All we have to do is to verify that the expansion property (1) in the definition of a conformal repeller holds true for a Cantor repeller. We exploit hyperbolic contraction (using Schwarz's lemma).

Each U_i is compactly contained in $V_{k(i)}$ for some $k(i) \in \{1, 2, \ldots, M\}$. The inclusion $U_i \to V_{k(i)}$ is a contraction of the corresponding hyperbolic metrics. In other words, there exists $\lambda_i > 1$ such that for all $z, w \in U_i$ we have

$$d_{U_i}(z, w) \geq \lambda_i d_{V_{k(i)}}(z, w) .$$

Let $\lambda = \min\{\lambda_i : 1 \leq i \leq m\} > 1$. If $z, w \in U_i$ and $f(z), f(w) \in V_j$ then, since f maps U_i conformally onto V_j, we have

$$d_{V_j}(f(z), f(w)) = d_{U_i}(z, w) \geq \lambda d_{V_{k(i)}}(z, w) .$$

Using the inequality in the above relations and an easy inductive argument (exercise), we see that if $z, w \in U_i$ are in the same connected component of $f^{-n}(V_j)$ for some j then

$$d_{V_j}(f^n(z), f^n(w)) \geq \lambda^n d_{V_{k(i)}}(z, w) . \tag{3.6}$$

Now suppose that z, w are points in $K_1 = J_f \cap V_{k(i)}$. Then $f^n(z), f^n(w) \in K_2 = J_f \cap V_j$. Since K_1 and K_2 are compact, the hyperbolic metrics $d_{V_{k(i)}}$ and d_{V_j} over K_1 and K_2 respectively are both comparable with the Euclidean metric. Hence there exists $C > 0$ depending on $V_{k(i)}, V_j, K_1$ and K_2 such that the inequality (3.6) translates into

$$|f^n(z) - f^n(w)| \geq C\lambda^n|z - w| . \tag{3.7}$$

Dividing both sides of (3.7) by $|z - w|$, fixing z and letting $w \to z$, we obtain $|(f^n)'(z)| \geq C\lambda^n$, as required. $\qquad\square$

The topological dynamics of a Cantor repeller is fairly easy to describe. First, we give a symbolic code for f in the limit set. We define the transition matrix of (f, J_f) as follows. Let $J_i = J_f \cap U_i$ for $i = 1, 2, \ldots, m$, and let the square matrix $A = (a_{ij})_{m \times m}$ be such that $a_{ij} = 1$ if $f(J_i) \supseteq J_j$ and $a_{ij} = 0$ otherwise. Now let $\Sigma_A \subseteq \{1, 2, \ldots, m\}^{\mathbb{N}}$ be the subspace of the space of infinite one-sided sequences in the m symbols $1, 2, \ldots, m$ defined by the condition that $x = (x_n)_{n \in \mathbb{N}} \in \Sigma_A$ if and only if $a_{x_n x_{n+1}} = 1$ for all n. We endow the set $\{1, 2, \ldots, m\}$ with the discrete topology and the cartesian product space $\{1, 2, \ldots, m\}^{\mathbb{N}}$ with the product topology, which makes it a compact space. Hence Σ_A is also a compact (Hausdorff, hence metrizable) space. It is *invariant* under the shift map $\sigma : \{1, 2, \ldots, m\}^{\mathbb{N}} \hookleftarrow$, given by $\sigma((x_n)_{n \in \mathbb{N}}) = (x_{n+1})_{n \in \mathbb{N}}$. The dynamical system (Σ_A, σ) is called the *subshift of finite type* associated with the transition matrix A.

As an example, figure 3.8 shows a Cantor repeller with transition matrix

$$A = \begin{pmatrix} 1 & 1 & 1 & 0 & 0 \\ 0 & 0 & 0 & 1 & 1 \\ 1 & 1 & 1 & 0 & 0 \\ 1 & 1 & 1 & 0 & 0 \\ 0 & 0 & 0 & 1 & 1 \end{pmatrix}.$$

Theorem 3.9.4 *The map f restricted to its limit set J_f is topologically conjugate to the subshift (Σ_A, σ) with transition matrix A.*

Proof Given $x \in J_f$, we know that for each $n \geq 0$ there exists a unique $i_n \in \{1, 2, \ldots, m\}$ such that $f^n(x) \in J_{i_n}$. We define the *itinerary* of x to be the sequence $\theta_x = i_0 i_1 \cdots i_n \cdots \in \{1, 2, \ldots, m\}^{\mathbb{N}}$. Note that $a_{i_n i_{n+1}} = 1$ for all $n \geq 0$, so that in fact $\theta_x \in \Sigma_A$. Hence we have a well-defined map $h : J_f \to \Sigma_A$ given by $h(x) = \theta_x$. We leave to the reader the task of proving that h is the desired conjugacy. $\qquad\square$

When filling in the blanks in the above proof, the reader will not fail to notice the following. Let us agree to call a finite sequence $i_0 i_1 \cdots i_n \in$

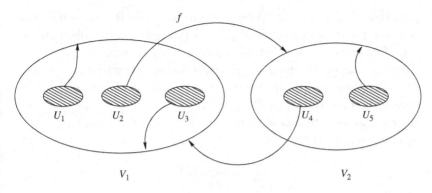

Fig. 3.8. A Cantor repeller.

$\{1, 2, \ldots, m\}^{n+1}$ *admissible* if $a_{i_k i_{k+1}} = 1$ for $k = 0, 1, \ldots, n - 1$. The connected components of $f^{-(n+1)}(V)$ can be labeled inductively by admissible sequences. Indeed, assuming that the components of $f^{-n}(V)$ have already been labeled, and given an admissible sequence $i_0 i_1 \cdots i_n$, let $U_{i_0 i_1 \cdots i_n}$ be the unique connected component of $f^{-(n+1)}(V)$ contained in $U_{i_0 i_1 \cdots i_{n-1}}$ with the property that $f(U_{i_0 i_1 \cdots i_n}) = U_{i_1 i_2 \cdots i_n}$. Each open set $U_{i_0 i_1 \cdots i_n}$ is a topological disk, because the open sets U_i at the base of the induction already are such. The reader will see that if $x \in J_f$ has itinerary $\theta_x = i_0 i_1 \cdots i_n \cdots$ then in fact

$$\{x\} = \bigcap_{n=0}^{\infty} U_{i_0 i_1 \cdots i_n} .$$

This follows because the diameters of the topological disks $U_{i_0 i_1 \cdots i_n}$ shrink to zero as $n \to \infty$. For a more quantitative estimate on the sizes of these disks, see lemma 3.9.10 below.

For each admissible sequence $i_0 i_1 \cdots i_n$, we define $J_{i_0 i_1 \cdots i_n} = J_f \cap U_{i_0 i_1 \cdots i_n}$ and call it the *cylinder* with prefix $i_0 i_1 \cdots i_n$. This clearly agrees with the image of the set of all sequences in Σ_A with prefix $i_0 i_1 \cdots i_n$ under h^{-1}, where h is the conjugacy built in theorem 3.9.4 above. The set of all such cylinders (with prefix given by a sequence of length $n+1$) will be denoted by $\mathcal{A}_n$.

3.9.2 Topological pressure

We are interested in the relationship between the dynamics of a Cantor repeller (f, J_f) and certain geometric features of the limit set J_f, such as its Hausdorff dimension. A beautiful discovery by R. Bowen [Bo] was that the Hausdorff dimension of J_f is in principle computable with the help of an ergodic-theoretic tool known as *topological pressure*.

Rather than giving the most general definition, we introduce the concept of topological pressure in the specific context of repellers. For a more general treatment, see [Wa] or [KH].

Let $\varphi : J_f \to \mathbb{R}$ be a continuous function, and let $B = J_{i_0 i_1 \cdots i_n} \in \mathcal{A}_n$ be a cylinder of J_f. We write

$$\varphi_B = \sup_{x \in B} \varphi(x) .$$

Let us also consider the Birkhoff sums of φ, namely

$$S_n \varphi = \sum_{j=0}^{n-1} \varphi \circ f^j .$$

Each of these sums is, in its own right, a continuous function on the limit set J_f.

Theorem 3.9.5 *For every continuous function* $\varphi : J_f \to \mathbb{R}$, *the limit*

$$P(\varphi) = \lim_{n \to \infty} \frac{1}{n} \log \sum_{B \in \mathcal{A}_n} e^{(S_n \varphi)_B} \tag{3.8}$$

exists.

Proof Let (p_n) be the sequence given by

$$p_n = \sum_{B \in \mathcal{A}_n} e^{(S_n \varphi)_B} .$$

We claim that (p_n) is sub-multiplicative, in the sense that $p_{m+n} \leq p_m p_n$ for all $m, n \geq 0$. To see why, note that for all $x \in J_f$ we have

$$S_{m+n} \varphi(x) = S_m \varphi(x) + S_n \varphi(f^m(x)) . \tag{3.9}$$

Given any cylinder $B \in \mathcal{A}_{m+n}$, we know that there exist cylinders $B' \in \mathcal{A}_m$ and $B'' \in \mathcal{A}_n$ such that $B = B' \cap f^{-m}(B'')$. Taking the supremum in (3.9) over all $x \in B$, we get

$$(S_{m+n} \varphi)_B \leq (S_m \varphi)_{B'} + (S_n \varphi)_{B''} .$$

Therefore we have

$$
\begin{aligned}
p_{m+m} &= \sum_{B \in \mathcal{A}_{m+n}} e^{(S_{m+n}\varphi)_B} \\
&\leq \sum_{B' \in \mathcal{A}_m} \sum_{B'' \in \mathcal{A}_n} e^{(S_m\varphi)_{B'} + (S_n\varphi)_{B''}} \\
&= \left(\sum_{B' \in \mathcal{A}_m} e^{(S_m\varphi)_{B'}} \right) \left(\sum_{B'' \in \mathcal{A}_n} e^{(S_n\varphi)_{B''}} \right) = p_m p_n .
\end{aligned}
$$

This proves our claim. Hence $(\log p_n)$ is a *sub-additive sequence*. The theorem follows, then, from a well-known lemma concerning sub-additive sequences (see exercise 3.20). $\qquad\square$

The limit (3.8) whose existence has been established by this theorem is called the *topological pressure* of φ. When $\varphi = 0$, such a limit agrees in fact with the *topological entropy* of $f|_{J_f}$, see [Wa].

3.9.3 Equilibrium states

Another notion from ergodic theory that will be needed below is that of the Gibbs or equilibrium measure. This concept has its origins in statistical mechanics, and its use in dynamical systems was pioneered by Sinai, Bowen and Ruelle (see [Zi] for a good exposition of the physical motivation behind this concept). Again, we restrict our discussion to the specific context at hand. Let us consider a (continuous) function $\varphi : J_f \to \mathbb{R}$.

Definition 3.9.6 *An equilibrium (or Gibbs) measure for φ is a Borel measure μ supported on J_f for which there exist constants $K \geq 1$ and $C > 0$ such that, for all cylinders $B \in \mathcal{A}_n$ and all $x \in B$, we have*

$$
\frac{1}{K} \leq \frac{\mu(B)}{e^{S_n\varphi(x)+Cn}} \leq K . \tag{3.10}
$$

Note that there exists *at most one* value of C for which (3.10) holds. The existence and uniqueness of an equilibrium measure for a given φ are not always guaranteed. A sufficient condition is to require that φ be Hölder continuous. Let us define the nth variation of φ to be

$$
\text{Var}_n \varphi = \max\{|\varphi(x) - \varphi(y)| : x, y \in B, B \in \mathcal{A}_n\} .
$$

We say that φ is Hölder continuous if the nth variation of φ decreases exponentially with n, i.e. if there exist constants $c > 0$ and $0 < \alpha < 1$ such that $\text{Var}_n \varphi \leq c\alpha^n$ for all n.

Theorem 3.9.7 *Let* $\varphi : J_f \to \mathbb{R}$ *be Hölder continuous. Then there exists a unique probability equilibrium measure for* (f, J_f, φ).

This is a special case of what is usually called the *Ruelle–Perron–Frobenius* theorem. We will not prove this theorem here. See [Zi, Chapter 4] or the article by M. Keane in [BKS] for a short proof. It turns out that, when φ is Hölder continuous, the unique equilibrium measure μ_φ satisfies (3.10) with $C = -P(\varphi)$, where $P(\varphi)$ is the topological pressure of φ.

3.9.4 Hausdorff dimension

We now digress a little to introduce the Hausdorff dimension. For more details and complete proofs of the assertions made here, see the standard reference [Fal].

First we define the so-called *Hausdorff outer measures* on $\mathbb{R}^n$. Given real numbers $s > 0$ and $\varepsilon > 0$ and any (Borel) set $E \subseteq \mathbb{R}^n$, let

$$\mu_s^\varepsilon E = \inf_{\mathcal{B}} \sum_{B \in \mathcal{B}} |B|^s ,$$

the infimum being taken over *all* coverings $\mathcal{B}$ of the set E by balls $B \in \mathcal{B}$ with diameter $|B| \leq \varepsilon$. Note that $\mu_s^\varepsilon(E)$ is, for fixed s and E, a non-increasing function of ε. Hence we can define $\mu_s(E) = \lim_{\varepsilon \to 0} \mu_s^\varepsilon(E)$. It is straightforward to prove that μ_s is an outer measure, for all $s > 0$. It is also an easy exercise to check that $\mu_s(E) = 0$ for all $E \subseteq \mathbb{R}^n$ when $s > n$.

Definition 3.9.8 *The Hausdorff dimension of a (Borel) set* $E \subseteq \mathbb{R}^n$ *is*

$$\dim_H E = \inf\{s > 0 : \mu_s(E) = 0\} .$$

In particular, the Hausdorff dimension of any $E \subseteq \mathbb{R}^n$ is always $\leq n$. One can show that if $d = \dim_H E$ then $\mu_s(E) = \infty$ if $s < d$ and $\mu_s(E) = 0$ if $s > d$. The Hausdorff dimension is *diffeomorphism invariant*: if $E \subseteq \mathbb{R}^n$ is a Borel set and $g : \mathbb{R}^n \to \mathbb{R}^n$ is a diffeomorphism then $\dim_H g(E) = \dim_H E$.

Calculating the exact value of the Hausdorff dimension of a set can be rather tricky. As a rule, good *upper* bounds are usually easier to obtain than good *lower* bounds. An important tool for good lower bounds is the following result.

Lemma 3.9.9 *Let $E \subseteq \mathbb{R}^n$ be a Borel set, and let μ be a (Borel) measure with support in E. Suppose there exist $s > 0$, $\varepsilon > 0$ and $C > 0$ such that $\mu(A) \leq C|A|^s$ for all measurable sets $A \subseteq E$ with $|A| \leq \varepsilon$. Then $\dim_H E \geq s$.*

Proof Let $\mathcal{B}$ be any (countable) covering of E by balls of diameter $\leq \delta < \varepsilon$. Then for all $B \in \mathcal{B}$ we have

$$\mu(B) = \mu(B \cap E) \leq C|B \cap E|^s \leq C|B|^s .$$

Therefore

$$\sum_{B \in \mathcal{B}} |B|^s \geq \frac{1}{C} \sum_{B \in \mathcal{B}} \mu(B) \geq \frac{1}{C} \mu \left(\bigcup_{B \in \mathcal{B}} B \right)$$
$$= \frac{1}{C} \mu(E) > 0$$

Taking the infimum over all such coverings, we get $\mu_s^\delta(E) \geq C^{-1} \mu(E) > 0$. Letting $\delta \to 0$, we deduce that $\mu_s(E) > 0$, and this means of course that $\dim_H E \geq s$. ☐

The above lemma is known in the literature as the *mass distribution principle* (see for instance [Fal, p. 60]). In some places it is called *Frostmann's lemma*, but this last name should be reserved for the more difficult result proved by Frostmann, namely the *converse* of the above lemma (which, however, will not be needed here). The mass distribution principle is extremely useful. As an exercise, the reader may employ it to calculate the Hausdorff dimension of the standard triadic Cantor set.

3.9.5 Bowen's formula

Let us now return to the problem of computing the Hausdorff dimension of our Cantor repeller J_f. Let us consider the function $\psi = -\log|f'|$ restricted to J_f. Note that the expansion property (1) defining a conformal repeller implies that, for all sufficiently large n, the Birkhoff sums $S_n \psi$ are *negative* everywhere in J_f. This fact will be used below. But first we need the following lemma, which estimates the sizes of cylinders of J_f in terms of the values of these Birkhoff sums on points of the cylinders.

Lemma 3.9.10 *There exists a constant $C > 0$ such that, for all cylinders $B \in \mathcal{A}_n$ and all points $x \in B$, we have*

$$C^{-1} e^{S_n \psi(x)} \leq |B| \leq C e^{S_n \psi(x)} .$$

Proof First note that the maximum diameter of cylinders in $\mathcal{A}_n$ goes to zero as $n \to \infty$, as one can see from the proof of theorem 3.9.4. This allows us to choose $n_0 \in \mathbb{N}$ sufficiently large and $r > 0$ sufficiently small for the following to hold:

(1) every cylinder $A \in \mathcal{A}_{n_0}$ is contained in the disk of radius $r/2$ about any $y \in A$;

(2) if $A \in \mathcal{A}_{n_0}$ and $V_A \in \{V_1, \ldots, V_M\}$ is such that $A \subset V_A$ then V_A contains the disk of radius r about any point of A.

From (1) and (2) it follows that, if $n > n_0$ and $B \in \mathcal{A}_n$ is such that $f^{n-n_0}(B) = A \in \mathcal{A}_{n_0}$, then the inverse branch g of f^{n-n_0} that maps A back onto B, being univalent in V_A, is *a fortiori* univalent in the disk of radius r about any point of A. Given $x \in B$, let $y = f^{n-n_0}(x)$ (hence $x = g(y)$). The disk D of radius $r/2$ about y is a convex set inside V_A whose distance to the boundary of V_A is $\geq r/2$. Let $y_1, y_2 \in A \subset D$ be such that $|g(y_1) - g(y_2)| = |B|$. Applying lemma 3.8.3 to this situation, we see that

$$|B| = |g(y_1) - g(y_2)| \leq K|g'(y)| \, |y_1 - y_2| .$$

But $|y_1 - y_2| \leq |A| < r/2$, whereas

$$|g'(y)| = \frac{1}{|(f^{n-n_0})'(x)|}$$

$$= \exp\left\{ -\sum_{j=0}^{n-n_0-1} \log|f'(f^j(x))| \right\} = e^{S_{n-n_0}\varphi(x)} .$$

Moreover, $S_{n-n_0}\varphi(x) = S_n\varphi(x) - S_{n_0}\varphi(f^{n-n_0}(x))$. Therefore we have

$$|B| \leq C_1 e^{S_n \varphi(x)}$$

for all $n > n_0$ and all $B \in \mathcal{A}_n$, where

$$C_1 = \frac{Kr}{2} \max_{z \in J_j} e^{-S_{n_0}\varphi(z)} .$$

To prove the required lower bound, take $z_1, z_2 \in A$ such that $|z_1 - z_2| = |A|$. Again applying lemma 3.8.3, this time using the lower estimate provided by that lemma, we get

$$|B| \geq |g(z_1) - g(z_2)| \geq \frac{1}{K}|g'(y)| |A| .$$

Hence we have

$$|B| \geq \frac{1}{K}|A|e^{S_{n-n_0}\varphi(x)} .$$

Now observe that $S_n\varphi(x) < S_{n-n_0}\varphi(x)$ because $\varphi = -\log|f'|$ is everywhere negative in J_f (recall that f is expanding there). Therefore

$$|B| \geq C_2 e^{S_n\varphi(x)} ,$$

where $C_2 = K^{-1}\min_{A\in\mathcal{A}_{n_0}} |A|$. This proves the lemma, provided that we take $C = \max\{C_1, C_2^{-1}\}$.

$\square$

This lemma tells us in particular that the sizes of cylinders in $\mathcal{A}_n$ decrease at an exponential rate as $n \to \infty$. In particular, $\varphi = -\log|f'|$ is Hölder continuous in J_f. Hence by theorem 3.9.7 there exists an equilibrium measure for φ. This fact will be used in the following theorem, due to Bowen, the culmination of our efforts in this section.

Theorem 3.9.11 (Bowen) *Let (f, J_f) be a Cantor repeller. Then the Hausdorff dimension of its limit set J_f is the unique real number t such that $P(-t\log|f'|) = 0$.*

Proof Let us first prove that a value of t with the stated property exists. We will write $\varphi = -\log|f'|$ as before. Recall that

$$P(t\varphi) = \lim_{n\to\infty} \frac{1}{n}\log p_n(t) ,$$

where

$$p_n(t) = \sum_{B\in\mathcal{A}_n} e^{t(S_n\varphi)_B} . \tag{3.11}$$

As observed before, $P(0)$ is equal to the topological entropy of (f, J_f) and this turns out to be a positive number, so $P(0) > 0$. However, because f is expanding in J_f, say $|f'| \geq \lambda > 1$ there, we have

$$\varphi = -\log|f'| \leq -\log\lambda < 0 .$$

Since $\mathcal{A}_n$ contains at most N^{n+1} cylinders (where N is the number of components U_i in the domain of f), we see from (3.11) that $p_n(t) \leq$

$N^{n+1}e^{-\lambda tn}$. Hence

$$\frac{1}{n}\log p_n(t) \ \leq \ \left(1+\frac{1}{n}\right)\log N - \lambda t \ ,$$

and this tells us that

$$P(t\varphi) \ \leq \ \log N - \lambda t \ .$$

This shows that $P(t\varphi) \to -\infty$ as $t \to \infty$. A similar sort of argument also shows that P is a decreasing function of t. Therefore there exists a unique value of t for which $P(t\varphi) = 0$. Let us denote this special value of t by δ. Note that $\delta > 0$.

Now we need to show that δ is the Hausdorff dimension of J_f. First we claim that $\dim_H J_f \leq \delta$. Let us use the coverings of J_f given by the $\mathcal{A}_n$ themselves. Given any $t > \delta$ and $\varepsilon > 0$, choose n sufficiently large that every $B \in \mathcal{A}_n$ has diameter less than ε. Applying lemma 3.9.10, we get

$$\mu_t^\varepsilon(J_f) \ \leq \ \sum_{B\in\mathcal{A}_n} |B|^t$$

$$\leq \ C^t \sum_{B\in\mathcal{A}_n} e^{t(S_n\varphi)_B} \ = \ C^t p_n(t) \ ,$$

where C is the constant in that lemma. From this, it follows that

$$\mu_t^\varepsilon(J_f) \ \leq \ C^t e^{n(P(t\varphi)-\epsilon_n)} \ , \tag{3.12}$$

where $\epsilon_n = P(t\varphi) - (\log p_n(t)/n)$ tends to zero as $n \to \infty$. Since $P(t\varphi) < 0$, the right-hand side of (3.12) also goes to zero as $n \to \infty$, and thus $\mu_t^\varepsilon(J_f) = 0$ for all $\varepsilon > 0$. Hence $\mu_t(J_f) = 0$ for all $t > \delta$, and this shows that $\dim_H(J_f) \leq \delta$ as claimed.

In order to reverse this inequality, we apply the mass distribution principle using the equilibrium measure μ for the potential $\delta\varphi$, whose existence is guaranteed by theorem 3.9.7. We need to check that μ satisfies the hypothesis of that theorem with $s = \delta$. In other words, we need to show that

$$\mu(D(x,r)) \ \leq \ Cr^\delta \tag{3.13}$$

for every disk of sufficiently small radius r centered at an arbitrary point $x \in J_f$. Given such a disk, let $n = n(r,x)$ be chosen such that

$$|(f^{n-1})'(x)| \ < \ r^{-1} \ \leq \ |(f^n)'(x)| \ .$$

Let $\mathcal{B} \subseteq \mathcal{A}_n$ be the set of all cylinders B in $\mathcal{A}_n$ such that $B \cap D(x,r) \neq \emptyset$. Then

$$\mu(D(x,r)) \leq \sum_{B \in \mathcal{B}} \mu(B) . \tag{3.14}$$

It is not difficult to see (exercise) that the number of elements of $\mathcal{B}$ is bounded by a constant independent of n. Moreover, since μ is an equilibrium measure for $\delta\varphi$ and $P(\delta\varphi) = 0$, we see from (3.10) that

$$\mu(B) \leq C_1 e^{\delta S_n \varphi(y)} , \tag{3.15}$$

for every cylinder $B \in \mathcal{A}_n$, where $y \in B$ is arbitrary. For cylinders in $\mathcal{B}$, we can in fact take $y \in B \cap D(x,r)$. By lemma 3.8.3 (applied to a suitable inverse branch of f^n), for such points y we have

$$e^{S_n \varphi(y)} = |(f^n)'(y)|^{-1} \leq C_2 |(f^n)'(x)|^{-1} .$$

Incorporating this information in (3.15) yields

$$\mu(B) \leq C_3 |(f^n)'(x)|^{-\delta} \leq C_3 r^\delta ,$$

by our choice of n. Using this last inequality in (3.14) we get (3.13). This shows that μ indeed satisfies the hypothesis of lemma 3.9.9 and therefore that $\dim_H J_f \geq \delta$. This completes the proof of Bowen's theorem. $\qquad\square$

Remark 3.9.12 *The above result would remain valid if we considered uniformly asymptotically conformal (u.a.c.) Cantor repellers instead of conformal ones (see [dFGH] for the definition of u.a.c. Cantor repellers and more).*

Remark 3.9.13 *Bowen's formula is still valid for other conformal repellers, such as Julia sets of expanding rational maps. In particular, it holds true in the case of a quadratic polynomial $f_c(z) = z^2 + c$ with $|c|$ small. In this case, the limit set – that is, the Julia set $J(f_c)$ – is a quasicircle, i.e. the image of a round circle under a quasiconformal map, see Chapter 4. For a proof of Bowen's formula covering such cases, see [Zi]. In [Rue], D. Ruelle proved an asymptotic formula for the Hausdorff dimension of $J(f_c)$ for $|c|$ near zero, namely*

$$\dim_H J(f_c) = 1 + \frac{|c|^2}{4\log 2} + o(|c|^2) .$$

The proof involves the study of the first two derivatives of the pressure function. The Hausdorff dimension of $J(f_c)$ for c in the main cardioid *of the Mandelbrot set is a real analytic function of c, and it attains its minimum value 1 at c = 0. Once again, see [Zi, Chapter 6].*

Exercises

3.1   Prove that two Möbius transformations commute if and only if they have the same fixed points.

3.2   Let $V \subset \mathbb{C}$ be a domain whose complement has at least two points, and let $\rho_V(z)\,|dz|$ be the Poincaré metric on V. Show that

$$\Delta(\log \rho_V) \;=\; \frac{\partial^2}{\partial x^2}(\log \rho_V) + \frac{\partial^2}{\partial y^2}(\log \rho_V) \;=\; \rho_V^2 \;.$$

This expresses the fact (see the next exercise) that the Poincaré metric has Gaussian curvature equal to -1.

3.3   Let $\rho(z)|dz|$ be any conformal metric on a domain $V \subseteq \mathbb{C}$. Show that the Gaussian curvature of this metric is given by

$$K_\rho \;=\; -\frac{\Delta(\log \rho)}{\rho^2} \;.$$

3.4   Let $f : \mathbb{D} \to \mathbb{D}$ be holomorphic, and let ρ be the Poincaré density of $\mathbb{D}$. Show that if $f(\mathbb{D})$ is compactly contained in $\mathbb{D}$ then ρ is strictly contracted by f, i.e. there exists $0 < \lambda < 1$ such that $f_*\rho \leq \lambda\rho$. (Here $f_*\rho(z) = \rho(f(z))|f'(z)|$ for all z in the unit disk.)

3.5   Let f be a rational map. Prove that $P(f^n) = P(f)$ for all $n \geq 1$.

3.6   A point $z \in \widehat{\mathbb{C}}$ is said to be *exceptional* for a rational map f if its grand orbit

$$[z] \;=\; \{w \in \widehat{\mathbb{C}} : \exists\, m, n \geq 0 \text{ such that } f^m(w) = f^n(z)\}$$

is finite.

(a) Show that if f has degree ≥ 2 then f has at most two exceptional points.

(b) Show that if f has only one exceptional point then f is conjugate to a polynomial.

(c) Show that if f has exactly two exceptional points then f is conjugate to $z \mapsto z^{\pm d}$ (where d is the degree of f).

3.7   Let $f : V \to \mathbb{C}$ be an analytic map, and let $z_0 \in V$ be a fixed point of f such that $\lambda = f'(z_0)$ satisfies $0 < |\lambda| < 1$. Show that f

is locally analytically conjugate to the linear map $z \mapsto \lambda z$, working through the following steps.

    (a) There exists a disk $D \subset V$ centered at z_0 such that $f(D) \subset D$.

    (b) Define $\varphi_n : D \to \mathbb{C}$ by

$$\varphi_n(z) = \frac{f^n(z)}{\lambda^n} .$$

    Show that $\{\varphi_n\}$ converges uniformly on compact subsets of D to an analytic map $\varphi : D \to \mathbb{C}$.

    (c) Show that φ satisfies $\varphi \circ f = \lambda \varphi$ everywhere in D and check also that $\varphi'(0) = 1$, so that φ is a local analytic diffeomorphism.

3.8   Prove that, in the previous exercise, the conjugacy of f to its linear part $z \mapsto \lambda z$ is (locally) unique up to multiplication by a non-zero scalar.

3.9   Using exercise 3.7, prove that if f is analytic and z_0 is an *expanding* fixed point, i.e. the multiplier $\lambda = f'(z_0)$ satisfies $|\lambda| > 1$, then f is locally analytically conjugate to $z \mapsto \lambda z$.

3.10   Let $g : V \to \mathbb{C}$ be analytic in a neighborhood V of the origin, and suppose that g has a zero of order $m \geq 2$ at 0. Show that there exists a map h, which is analytic in the neighborhood of 0, such that $g(z) = (h(z))^m$.

3.11   Let f be analytic and let z_0 be a super-attracting fixed point of f, so that in the neighborhood of z_0 we have

$$f(z) = z_0 + a(z - z_0)^k + \cdots ,$$

where $k \geq 2$ and $a \neq 0$. Prove that f is locally analytically conjugate to $z \mapsto z^k$, working through the following steps.

    (a) First show that, by a suitable (affine) linear conjugation, we may assume that $z_0 = 0$ and $a = 1$.

    (b) Prove that, for $|z|$ sufficiently small, we have $|f^n(z)| \leq (C|z|)^{k^n}$ for all n, where $C > 0$ is a constant.

    (c) Show that, in a small but fixed neighborhood of 0, for each $n \geq 1$ there exists an analytic function φ_n such that

$$\varphi_n(z)^{k^n} = f^n(z)$$

    for all z in that neighborhood (use exercise 3.10). Also, check that $\varphi_n'(0) = 1$ for all n.

(d) Prove that

$$\left| \frac{\varphi_{n+1}(z)}{\varphi_n(z)} \right| = 1 + O(k^{-n})$$

for all $n \geq 1$ and all z in a small disk D centered at 0 (D is independent of n).

(e) Deduce from (d) that $\prod_{n=1}^{\infty} \varphi_{n+1}/\varphi_n$ converges uniformly in D to an analytic map $\varphi : D \to \mathbb{C}$ satisfying $\varphi'(0) = 1$.

(f) Verify that $\varphi \circ f(z) = (\varphi(z))^k$ for all $z \in D$, so that φ is indeed a local conjugacy of f to the power map $z \mapsto z^k$.

3.12   Prove that the conjugacy φ obtained in exercise 3.11 is unique up to multiplication by a $(k-1)$th root of unity.

3.13   Let $f, g : \widehat{\mathbb{C}} \to \widehat{\mathbb{C}}$ be rational maps, both having degrees ≥ 2. Prove that if $f \circ g = g \circ f$ then $J(f) = J(g)$.

3.14   Let $f : \widehat{\mathbb{C}} \to \widehat{\mathbb{C}}$ be a rational map of degree ≥ 2, and let

$$\Gamma = \{\gamma \in \text{Möb}(\widehat{\mathbb{C}}) : \gamma \circ f = f \circ \gamma\} \ .$$

Show that Γ is a finite group. (*Hint* Look at periodic points of f whose minimal periods are prime.)

3.15   Let $f : \widehat{\mathbb{C}} \to \widehat{\mathbb{C}}$ be the polynomial $f(z) = z^2 - 2$.

(a) Show that both the interval $[-2, 2] \subset \widehat{\mathbb{C}}$ and its complement $\Omega = \widehat{\mathbb{C}} \setminus [-2, 2]$ are completely invariant under f.

(b) Show that (f^n) is normal in Ω; this implies that $J(f) \subseteq [-2, 2]$.

(c) Using Vitali's theorem (see exercise 2.7), show that f^n converges to ∞ uniformly on compact sets in Ω, and deduce from this that $J(f) = [-2, 2]$.

3.16   Show that a polynomial map $f : \widehat{\mathbb{C}} \to \widehat{\mathbb{C}}$ cannot have Herman rings. (*Hint* Apply the maximum principle.)

3.17   This exercise characterizes those rational maps $f : \widehat{\mathbb{C}} \to \widehat{\mathbb{C}}$ whose Julia sets are the whole Riemann sphere.

(a) Let $\{U_n\}_{n \geq 1}$ be a countable base for the topology on $\widehat{\mathbb{C}}$, and let $V \subseteq \widehat{\mathbb{C}}$ be the set of all points whose forward orbits are dense in $\widehat{\mathbb{C}}$. Show that

$$D = \bigcap_{n \geq 1} \bigcup_{k \geq 1} f^{-k}(U_n) \ .$$

(b) Using (a) and Baire's theorem, show that if $J(f) = \widehat{\mathbb{C}}$ then $V \neq \emptyset$.

(c) Conversely, suppose that the forward orbit of $z \in \widehat{\mathbb{C}}$ is dense but the Fatou set $F(f)$ is non-empty.

(1) Show that $z \in F(f)$.

(2) If W is the connected component of $F(f)$ containing z, show that $F(f) = \cup_{i=0}^{N-1} f^i(W)$ for some $N \geq 1$.

(3) Show that this violates theorem 3.7.13.

(d) Deduce from the above that $J(f) = \widehat{\mathbb{C}}$ if and only if there exists $z \in \widehat{\mathbb{C}}$ whose forward orbit under f is dense in $\widehat{\mathbb{C}}$.

3.18 *Cremer examples.* This exercise outlines a proof of the existence of irrationally indifferent fixed points of analytic maps that are *not* linearizable. Let $f(z) = \lambda z + z^d$, where $d \geq 2$ and $\lambda = e^{2\pi i\theta}$ (with θ irrational).

(a) Show that there are $d^n - 1$ non-zero solutions to $f^n(z) = z$ (including multiplicities). Labeling these periodic points $z_1, z_2, \ldots, z_{d^n-1}$, show that

$$\prod_{j=1}^{d^n-1} |z_j| = |\lambda^n - 1| .$$

(b) Using (a), show that if λ satisfies

$$\liminf_{n\to\infty} |\lambda^n - 1|^{1/d^n} = 0 \qquad (\text{E3.1})$$

then there is a sequence of periodic points of f converging to zero. Deduce that f is not linearizable at zero.

(c) Working with the continued fraction expansion of θ, show that one can choose λ so that (E3.1) is satisfied.

(d) Prove that the set of θ's satisfying (E3.1) is dense in $\mathbb{R}$.

3.19 Construct a Cantor repeller whose limit set is the standard middle-thirds Cantor set.

3.20 Let $(a_n)_{n\in\mathbb{N}}$ be a sequence of non-negative real numbers. Show that if the sequence is sub-additive, i.e. if $a_{m+n} \leq a_m + a_n$ for all m, n, then $\lim_{n\infty} a_n/n$ exists.

3.21 Show with the help of the mass distribution principle that the Hausdorff dimension of the standard middle-thirds Cantor set is $\log 2/\log 3$.

4

The measurable Riemann mapping theorem

One of the most important modern tools in complex dynamics is the notion of quasiconformal homeomorphism. In complex dynamics, we study holomorphic systems and try to classify them up to holomorphic or conformal equivalence. This is sometimes difficult to achieve because conformal maps are a bit too rigid. So it is useful to relax the notion of conformal equivalence to a more flexible one. Quasiconformal equivalence is such a notion. Using quasiconformal homeomorphisms, we can perform surgeries on conformal maps. The measurable Riemann mapping theorem is the major tool that allows us to recover a new holomorphic dynamical system after the surgery. In this chapter we introduce the elementary theory of quasiconformal maps and prove the measurable Riemann mapping theorem. Then we present some dynamical applications, including Sullivan's solution to Fatou's wandering-domains problem.

4.1 Quasiconformal diffeomorphisms

Let $f: U \to V$ be an orientation-preserving C^1 diffeomorphism between open domains in the plane. Since the Jacobian of f, namely $J(z) = |\partial f(z)|^2 - |\bar{\partial} f(z)|^2$, is positive, we have $|\partial f(z)| \neq 0$. Therefore we can define the *complex dilatation* of f at z to be the number

$$\mu(z) = \frac{\bar{\partial} f(z)}{\partial f(z)} \ .$$

Note that $\mu(z) < 1$ for all z. If the absolute value of the complex dilatation of f is bounded away from 1 throughout U, we say that f is a *quasiconformal* diffeomorphism. More precisely, given a real number $K \geq 1$, we say that f is *K-quasiconformal* if its complex dilatation satisfies

$$|\mu(z)| \le \frac{K-1}{K+1}$$

for every $z \in U$. Geometrically, this means that, for every $z \in U$, the derivative of f at z maps each ellipse centered at the origin with eccentricity

$$\frac{1+|\mu(z)|}{1-|\mu(z)|} \le K$$

and whose major axis makes an angle $(\arg \mu(z))/2$ with the real axis onto a circle. The function μ is also called the *Beltrami coefficient* of f; the positive real number $K(z) = (1 + |\mu(z)|)/(1 - |\mu(z)|)$ is the *conformal distortion* of f at the point z. It is clear that a C^1 diffeomorphism that is 1-quasiconformal is in fact a holomorphic diffeomorphism. If f is a composition of a conformal diffeomorphism with a K-quasiconformal diffeomorphism then f is also K-quasiconformal. The inverse of a K-quasiconformal diffeomorphism is again K-quasiconformal. The composition of a K-quasiconformal diffeomorphism with a K'-quasiconformal diffeomorphism is KK'-quasiconformal. These facts are left as exercises. Here is another fact that we will need below. If f is a quasiconformal diffeomorphism as above then the norm of the derivative of f at each point is bounded by the square root of the product of the conformal distortion and the Jacobian; in other words,

$$\|Df(z)\| \le \sqrt{K(z)J(z)} \ .$$

This follows because, as the reader can check, the Jacobian determinant of f at each point is the product of the largest expansion of the derivative by the smallest, and the quasiconformal distortion is the ratio of these quantities.

The proof of the theorem below is known as Grötzsch's argument.

Theorem 4.1.1 (Grötzsch) *Let $\phi\colon R_1 \to R_2$ be a K-quasiconformal diffeomorphism between two ring domains of finite modulus. Then*

$$\frac{1}{K} \bmod R_1 \le \bmod R_2 \le K \bmod R_1 \ .$$

Proof We may suppose that the ring domains are round annuli, say $R_1 = A_r$ and $R_2 = A_R$. Let C_r be the cylinder $S^1 \times (0, \log r)$, and let C_R be similarly defined. Since the mapping $C_r \to A_r$ defined by $(e^{i\theta}, t) \mapsto te^{i\theta}$ is conformal, and similarly for C_R and A_r, we have that

ϕ induces a K-quasiconformal diffeomorphism $\varphi : C_r \to C_R$. For each $0 \le \theta \le 2\pi$, let $\gamma_\theta(t) = \varphi(e^{i\theta}, t)$. Since this curve joins the two boundary components of C_R, its length is greater than or equal to $\log R$, in other words,

$$\log R \le \int_0^{\log r} \left| \frac{\partial \varphi}{\partial t} \right| dt .$$

If we integrate this inequality with respect to θ and use the fact that $|\partial\varphi/\partial t| \le \sqrt{K_\varphi J_\varphi}$, we get

$$2\pi \log R \quad \le \quad \int_0^{2\pi} \int_0^{\log r} \left| \frac{\partial \varphi}{\partial t} \right| d\theta dt$$

$$\le \quad \iint_{C_r} \sqrt{K_\varphi J_\varphi} \, d\theta dt .$$

Now the Cauchy–Schwarz inequality applied to this last integral yields

$$\iint_{C_r} \sqrt{K_\varphi J_\varphi} \, d\theta dt \le \left(\iint_{C_r} K_\varphi \, d\theta dt \right)^{1/2} \left(\iint_{C_r} J_\varphi \, d\theta dt \right)^{1/2} .$$

Therefore, since $K_\varphi \le K$ everywhere, we have

$$(2\pi \log R)^2 \le K \iint_{C_r} d\theta dt \iint_{C_r} J_\varphi \, d\theta dt \le K(2\pi \log r)(2\pi \log R) .$$

Since $\mathrm{mod}\, A_R = (\log R)/2\pi$, this last inequality implies that

$$\mathrm{mod}\, A_R \le K \, \mathrm{mod}\, A_r.$$

This proves the lower inequality in the theorem statement. The upper inequality is obtained in the same way using the inverse mapping. $\quad\square$

The same computation illustrates how quasiconformal diffeomorphisms deform the modulus of another type of domain, the rectangles. For $a, b > 0$ we denote by $R(0, a, a + bi, bi) = \{z = x + iy \in \mathbb{C}; 0 \le x \le a, 0 \le y \le b\}$ the rectangle with vertices $0, a, a + bi, bi$. The modulus of the rectangle $R = R(0, a, a + bi, bi)$ is the number $\mathrm{mod}\, R = a/b$.

Theorem 4.1.2 *If $f : R(0, a, a+bi, bi) \to R(0, a', a' + b'i, b'i)$ is a homeomorphism that maps vertices to vertices in the given order and if f is a K-quasiconformal diffeomorphism in the interior of the rectangles then*

$$\frac{1}{K} \left(\frac{a}{b} \right) \le \frac{a'}{b'} \le K \left(\frac{a}{b} \right) .$$

Proof Grötzsch's argument used above applies here also, *mutatis mutandis*. □

4.2 Extremal length and moduli of quadrilaterals

We will introduce here an important conformal invariant that is very useful for obtaining estimates of the moduli of quadrilaterals and rings.

Let Γ be a family of piecewise C^1 curves in the plane. We will assign to Γ a non-negative real number $\lambda(\Gamma)$ which will be a conformal invariant, in the sense that if f is a conformal mapping whose domain contains all the curves of the family then $\lambda(f(\Gamma)) = \lambda(\Gamma)$. If ρ is a non-negative measurable function, we define the ρ-length of a piecewise C^1 curve γ by

$$l_\rho(\gamma) = \int_\gamma \rho |dz| = \int \rho(\gamma(t)) |\gamma'(t)| dt$$

if $t \mapsto \rho(\gamma(t))$ is measurable and $l_\rho(\gamma) = \infty$ otherwise. Next, we define the ρ-length of the family Γ, namely

$$L_\rho(\Gamma) = \inf_{\gamma \in \Gamma} l_\rho(\gamma) .$$

Finally the *extremal length* of the family is defined as

$$\lambda(\Gamma) = \sup_\rho \frac{(L_\rho(\Gamma))^2}{A_\rho} ,$$

where the supremum is over the set of all non-negative measurable ρ values with finite area,

$$A_\rho = \iint_{\mathbb{C}} \rho^2 \, dx dy < \infty .$$

If there exists a conformal metric $\rho |dz|$ that realizes the extremal length, it is called an *extremal metric* of the family.

Theorem 4.2.1 *If* $f \colon U \to \hat{V}$ *is a K-quasiconformal diffeomorphism and Γ is a family of piecewise C^1 curves in U then*

$$\frac{1}{K}\lambda(\Gamma) \leq \lambda(f(\Gamma)) \leq K\lambda(\Gamma).$$

In particular, the extremal length is a conformal invariant.

Proof To estimate the extremal length of Γ, we may restrict our attention to conformal metrics that vanish outside U. For each such metric $\rho\,|dz|$, let $\hat{\rho}\,|dw|$ be the conformal metric defined by

$$\hat{\rho}(w) = \frac{\rho(z)}{|\partial f(z)| - |\bar{\partial} f(z)|} \; ,$$

where $w = f(z)$ (we set $\hat{\rho}(w) = 0$ for all $w \notin V$). If $\gamma \in \Gamma$ and $\hat{\gamma} = f \circ \gamma \in f(\Gamma)$ then, since

$$\hat{\gamma}'(t) = \partial f(z)\gamma'(t) + \bar{\partial} f(z)\overline{\gamma'(t)} \; ,$$

we have $L_\rho(\gamma) \le L_{\hat{\rho}}(\hat{\gamma})$. However, since the Jacobian of f at z is equal to $|\partial f(z)|^2 - |\bar{\partial} f(z)|^2$ we have

$$\iint_V \hat{\rho}^2(w)\,du\,dv \;=\; \iint_U \hat{\rho}^2 \circ f(z)(|\partial f(z)|^2 - |\bar{\partial} f(z)|^2)\,dx\,dy$$

$$=\; \int_U \rho^2(z)\frac{|\partial f(z)| + |\bar{\partial} f(z)|}{|\partial f(z)| - |\bar{\partial} f(z)|}\,dx\,dy \; .$$

Here we have written $w = u + iv$ in the first integral. Note that the ratio appearing in the last integral is the conformal distortion $K(z) \le K$. Therefore

$$\iint_V \hat{\rho}^2\,du\,dv \le K \iint_U \rho^2\,dx\,dy \; .$$

In other words, $A_{\hat{\rho}} \le KA_\rho$. This shows that $\lambda(f(\Gamma)) \ge K^{-1}\lambda(\Gamma)$. We may obtain the other inequality by the same argument using the inverse of f. $\square$

Example 4.2.2   *The modulus of a rectangle as an extremal length.* Let $R = R(0, a, a + bi, bi)$ be the rectangle with vertices $0, a, a + bi, bi$. Let Γ be the family of piecewise C^1 curves in R connecting the two vertical sides of R. The extremal length of the family Γ is equal to a/b, and $\rho_e = 1$ in R and 0 in the complement of R is the unique extremal metric (up to multiplication by a positive constant). Indeed, for this conformal metric, the area of the rectangle is ab and the smallest curve is a horizontal segment with length a. Hence we have

$$\frac{(L_{\rho_e}(\Gamma))^2}{A_{\rho_e}} \;=\; \frac{a^2}{ab} = \frac{a}{b} \; .$$

It remains to prove that

$$\frac{(L_\rho(\Gamma))^2}{A_\rho} \;<\; \frac{a}{b} \tag{4.1}$$

for any other conformal metric $\rho|dz|$ if ρ is not a positive multiple of ρ_e. This is clear if $L_\rho(\Gamma) = 0$. Otherwise, we can set $L_\rho(\Gamma)) = a$ by multiplying ρ by a positive constant. In particular, the ρ-length of each horizontal segment is greater than or equal to a, that is to say

$$\int_0^a (\rho(z) - 1)\, dx \geq 0 .$$

Integrating with respect to y we get

$$\iint_R (\rho - 1)\, dxdy \geq 0 .$$

However, if the Jacobian of f at z is equal to ρ, and ρ is not identically equal to unity in R, we have

$$\begin{aligned}
0 < \iint_R (\rho - 1)^2 &= \iint_R \rho^2 - 2\iint_R \rho + \iint_R 1 \\
&\leq \iint_R \rho^2 - \iint_R 1 .
\end{aligned}$$

This shows that

$$A_\rho = \iint_R \rho^2 > \iint_R 1 = ab ,$$

from which (4.1) follows. Therefore the extremal length of the family Γ is equal to a/b and $|dz|$ is the extremal metric. If Γ^* is the set of curves in the rectangle connecting the two horizontal sides then the same argument proves that the extremal length of Γ^* is equal to b/a. Thus the product of the extremal lengths of both families is equal to unity.

Example 4.2.3 *Modulus of an annulus as an extremal length.* Let Γ be the set of piecewise C^1 curves in the annulus A_R connecting its two boundary components. An argument similar to the previous example shows that the extremal length of the family Γ is equal to $(2\pi)^{-1}\log R$ and that the extremal metric is $\rho|dz|$ where $\rho(z) = |z|^{-1}$. It follows that, for any topological annulus A, the extremal length of the family of piecewise C^1 curves connecting the boundary components of A is equal to the modulus of A and, up to a multiplicative constant, there is a unique extremal metric. Again, if Γ^* is the set of closed curves that separate the two components of $\mathbb{C} \setminus A_R$ then the extremal length of Γ^* is the inverse of the extremal length of Γ, as a similar computation shows.

Definition 4.2.4 (Quadrilateral) *A quadrilateral* $Q(z_1, z_2, z_3, z_4)$ *is a Jordan domain with four distinct points* z_1, z_2, z_3, z_4 *marked in the boundary of* Q *in the counterclockwise sense.*

The positively oriented arcs in the boundary connecting z_1 to z_2 and z_3 to z_4 are called *horizontal sides*, whereas the arcs connecting z_2 to z_3 and z_4 to z_1 are called *vertical sides*. The *modulus*, mod Q, of the quadrilateral $Q = Q(z_1, z_2, z_3, z_4)$ is defined as the extremal length of the family of piecewise C^1 curves connecting the vertical sides (z_1, z_4) and (z_2, z_3).

Proposition 4.2.5 (Canonical map of quadrilaterals) *There exists a homeomorphism of the closure of the quadrilateral Q onto the closure of a rectangle R that preserves the vertices and is conformal in the interior of the quadrilateral.*

Proof By theorem 3.4.4, the Riemann mapping of any quadrilateral extends continuously to a homeomorphism of the closure of the quadrilateral, and we can normalize it in such a way that z_1 is mapped to 1, z_2 is mapped to i and z_3 is mapped to -1. The fourth vertex z_4 is mapped to some point with a negative imaginary part. Let $f_a : R_a = R(0, a, a + i, i) \to \mathbb{D}$ be the Riemann mapping with the above normalization. We claim that $\theta(a) = \mathrm{Re}\, f_a(i)$ is a continuous function of a whose range is the whole interval between 1 and -1. This clearly proves the proposition. Let us prove the continuity of θ at a point a_0. Take any sequence a_n converging to a_0. Choose a basepoint z_0 in the rectangle R_{a_0}; we may assume that z_0 belongs to R_{a_n} for all n. For each n, choose an affine Möbius transformation T_n such that $T_n(z_0) = z_0$, $T_n'(z_0) > 0$, $\hat{R}_{a_n} = T_n(R_{a_n})$ contains R_{a_0} in its interior, $\hat{R}_{a_{n+1}} \subset \hat{R}_{a_n}$ and T_n converges to the identity as $n \to \infty$. If $\phi_n : R_{a_n} \to \mathbb{D}$ is the Riemann mapping that maps z_0 to 0 and has a positive derivative at z_0 then $\hat{\phi}_n = \phi_n \circ T_n^{-1}$ is the Riemann mapping of $\hat{R}_n$ with the same normalization. By theorem 3.4.5, the inverse of $\hat{\phi}_n$ converges to the inverse of ϕ_0 uniformly in the closure of $\mathbb{D}$. Thus, the same holds for the inverse of ϕ_n, since T_n converges uniformly to the identity. Therefore the image of the four vertices $0, a_n, a_n + i, i$ by ϕ_n converges to the image of the vertices $0, a_0, a_0 + i, i$ by ϕ_0. Hence the sequence of Möbius transformations that maps the image of the first three vertices into $1, i, -1$ converges. This proves the continuity of θ at a_0. If $a_n \to \infty$ then $\theta(a_n) \to -1$; otherwise, taking a subsequence that is bounded away from -1, we would have that the Euclidean length of any curve connecting the vertical sides in $\mathbb{D}$ would be bounded away from zero. Hence, the extremal length of this family of curves would be bounded away from zero. But this is the same as the extremal length of the family of curves connecting the vertical sides of R_{a_n}, which clearly converges to zero. Similarly we can prove that $\theta(a) \to 1$ if $a \to 0$. $\qquad\square$

Therefore, if Q is a quadrilateral of modulus M there exists a homeomorphism f of the closure of Q onto the closure of the rectangle $R(0, M, M + i, i)$ that preserves the vertices and is holomorphic in the interior of Q. This map is called the *canonical map* of the quadrilateral Q. Hence there is a unique extremal metric, up to a multiplicative constant, for the family of curves connecting the vertical sides of Q, which is given by $\rho |dz|$ where $\rho(z) = |f'(z)|$.

Corollary 4.2.6 *The modulus of the quadrilateral* $Q(z_1, z_2, z_3, z_4)$ *is the inverse of the modulus of* $Q(z_2, z_3, z_4, z_1)$.

Proof The statement is clear for rectangles; see example 4.2.2. $\qquad\square$

Proposition 4.2.7 *Let Q be a quadrilateral. Consider a curve that connects the horizontal sides of Q and decomposes Q into two quadrilaterals Q_1 and Q_2. Then*

$$\mathrm{mod}\, Q \geq \mathrm{mod}\, Q_1 + \mathrm{mod}\, Q_2$$

and equality occurs if and only if Q_1 and Q_2 are mapped into rectangles by the canonical map of Q.

Proof We may suppose that Q is the rectangle $R(0, M, M + i, i)$. Let $f_i \colon Q_j \to R(0, M_j, M_j + i, i)$ be the canonical maps, $j = 1, 2$. Consider the conformal metric $\rho |dz|$ on Q, where $\rho(z) = \rho_j(z) = |f_j'(z)|$ if $z \in Q_j$ and is equal to zero otherwise. Let Γ_j be the family of curves connecting the vertical sides of Q_j and Γ the family connecting the vertical sides of Q. Clearly, $A_\rho(Q) = A_\rho(Q_1) + A_\rho(Q_2)$, as well as $\lambda(\Gamma_j) = M_j = A_\rho(Q_j)$. If $\gamma \in \Gamma$ then its ρ-length satisfies $l_\rho(\gamma) \geq L_\rho(\Gamma_1) + L_\rho(\Gamma_2)$. Therefore,

$$M = \lambda(\Gamma) \geq \frac{(L_\rho(\Gamma))^2}{A_\rho(Q)} \geq \frac{(\lambda_\rho(\Gamma_1) + \lambda_\rho(\Gamma_2))^2}{A_\rho(Q_1) + A_\rho(Q_2)}$$
$$= A_\rho(Q_1) + A_\rho(Q_2)$$
$$= \lambda(\Gamma_1) + \lambda(\Gamma_2) = M_1 + M_2 \,.$$

However, if we have equality then the first inequality above would be an equality and therefore $\rho|dz|$ would be the extremal metric of Q and so identically equal to unity. Hence f_1 is the identity, which implies that Q_1 is also a rectangle.   □

From this proposition and induction we get the following.

Corollary 4.2.8 *Let Q be a rectangle. Consider $n-1$ disjoint curves in Q connecting its horizontal sides and decomposing Q into quadrilaterals $Q_1, Q_2, \ldots Q_n$. Then*

$$\mod Q \geq \sum_{j=1}^{n} \mod Q_j \ ,$$

and the equality holds if and only if all the Q_j are rectangles.

The same proof gives a similar result concerning ring domains.

Proposition 4.2.9 *Let A be a ring domain. Let $A_1, \ldots, A_n \subset A$ be disjoint ring domains such that each A_j separates the two connected components of the complement of A. Then*

$$\mod A \geq \sum_{j=1}^{n} \mod A_j$$

and the equality holds if and only if the images of the A_i are mapped by the canonical map of A onto concentric round annuli and their complement in A is the union of $n-1$ round circles with center at the origin.

The corollary below was used by Yoccoz in the proof of his rigidity theorem for quadratic polynomials, which we will discuss in Chapter 5.

Corollary 4.2.10 *Let $A \subset \mathbb{C}$ be a bounded annulus. Suppose that there exist an infinite number of disjoint annuli $A_1, A_2, \ldots \subset A$ such that each A_i separates the components of the complement of A and*

$$\sum_{j=1}^{\infty} \mod A_j = \infty \ .$$

Then the bounded connected component of $\widehat{\mathbb{C}} \setminus A$ is a single point.

Proof If the bounded connected component of the complement of A were not a single point then the modulus of A would be finite, a contradiction.   □

4.3 Quasiconformal homeomorphisms

The notion of the quasiconformality of a map can be formulated in a way that makes it totally independent of any smoothness of the map. Thus we can talk about quasiconformal *homeomorphisms*.

Definition 4.3.1 (Quasiconformal homeomorphisms) *Let $K \geq 1$. An orientation-preserving homeomorphism $f \colon U \to V$ between domains of the Riemann sphere is K-quasiconformal if, for any quadrilateral Q whose closure is contained in U, we have*

$$\frac{1}{K} \operatorname{mod} Q \leq \operatorname{mod} f(Q) \leq K \operatorname{mod} Q \, .$$

If follows in particular that any K-quasiconformal diffeomorphism is a quasiconformal homeomorphism. Some immediate consequences of the definition are as follows.

(i) The inverse of a K-quasiconformal homeomorphism is a K-quasiconformal homeomorphism.

(ii) The composition of a K_1-quasiconformal homeomorphism with a K_2-quasiconformal homeomorphism is $K_1 K_2$-quasiconformal.

Next we will show, following Ahlfors [A1], that quasiconformality is a local property and that every 1-quasiconformal homeomorphism is conformal. We say that a homeomorphism is *locally K-quasiconformal* if every point has a neighborhood such that the restriction of the homeomorphism to this neighborhood is K-quasiconformal.

Theorem 4.3.2 *If $f \colon U \to V$ is a locally K-quasiconformal homeomorphism then f is K-quasiconformal.*

Proof Let Q be a quadrilateral of modulus M whose closure is contained in U and let M' be the modulus of its image Q'. Now we decompose Q into n thin quadrilaterals $Q_1, \ldots, Q_n$ by vertical lines in the canonical coordinates of Q. Hence, by corollary 4.2.8, the modulus of Q is equal to the sum of the moduli M_j of the Q_j. Let Q'_j be the image of Q_j. Then $M' \geq \sum_j M'_j$, where M'_j is the modulus of Q'_j. Next we decompose each Q'_j into small quadrilaterals $Q'_{j,k}$ using horizontal segments in the canonical coordinates of Q'_j. Hence

$$\frac{1}{M'_j} = \sum_k \frac{1}{M'_{j,k}} \, ,$$

where $M'_{j,k}$ is the modulus of $Q'_{j,k}$. If $M_{j,k}$ is the modulus of $Q_{j,k} = f^{-1}(Q'_{j,k})$ then we have

$$\frac{1}{M_j} \leq \sum_k \frac{1}{M_{j,k}} \, .$$

Finally, if both decompositions are fine enough, we have that $M_{j,k} \leq KM'_{j,k}$ because f is locally K-quasiconformal. Combining all these equalities and inequalities, we get $M \leq KM'$. The other inequality is proved similarly using the inverse of f. □

Theorem 4.3.3 *If $f \colon U \to V$ is 1-quasiconformal then f is conformal.*

Proof Let Q be a quadrilateral whose closure is contained in U and let Q' be its image. Composing f on the left with the canonical map of Q' and on the right with the inverse of the canonical map of Q, we get a 1-quasiconformal homeomorphism $g \colon R(0, M, M + i, i) \to R(0, M, M + i, i)$. It suffices to prove that g is the identity. Indeed, let $z \in R = R(0, M, M + i, i)$ and decompose the rectangle R into two subrectangles R_1, R_2 by the vertical line through z. Since g is 1-quasiconformal, we have $M = M_1 + M_2$, $M'_1 = M_1$, $M'_2 = M_2$ and $M = M'$. Thus $M' = M'_1 + M'_2$ and by proposition 4.2.7 the image R'_1 of R_1 is a rectangle having the same modulus as R_1, so it must coincide with R_1. Hence the vertical segment through z is mapped into itself. Similarly g maps the horizontal segment through z into itself. Hence $g(z) = z$, and since z is arbitrary, g is the identity and the theorem is proved. □

Corollary 4.3.4 *Let $f \colon U \to V$ be a homeomorphism and $\hat{f} \colon \mathbb{D} \to \mathbb{D}$ be a lift of f to the holomorphic universal covering spaces of U and V. Then f is a K-quasiconformal homeomorphism if and only if $\hat{f}$ is a K-quasiconformal homeomorphism.*

It is also possible to define quasiconformality using the distortion of the moduli of annuli instead of quadrilaterals. We refer to [LV, pp. 38–39] for the proof of the following.

Theorem 4.3.5 *An orientation-preserving homeomorphism is K-quasiconformal if and only if for every topological annulus A whose closure is contained in the domain we have*

$$\frac{1}{K} \operatorname{mod} A \leq \operatorname{mod} f(A) \leq K \operatorname{mod} A \, .$$

This theorem follows because we can use quadrilaterals to estimate the modulus of an annulus and annuli to estimate the moduli of quadrilaterals. From now on we will mainly consider quasiconformal homeomorphisms of the disk.

Next we will describe the distortion of quasiconformal homeomorphisms of the disk with respect to the distance induced by the hyperbolic metric.

Lemma 4.3.6 *There exists a strictly monotone, continuous and surjective function* $m \colon \mathbb{R}^+ \to \mathbb{R}^+$ *such that, given any two points* $p, q \in \mathbb{D}$, *there exists an annulus* A *and a degree-2 holomorphic covering map* $\phi_{p,q} \colon A \to \mathbb{D} \setminus \{p, q\}$ *such that the modulus of* A *is equal to* $m(d(p, q))$, *where* $d(p, q)$ *is the distance between* p *and* q *in the hyperbolic metric.*

Proof Consider the rational map $f \colon \widehat{\mathbb{C}} \to \widehat{\mathbb{C}}$ given by $f(z) = z + z^{-1}$. This map has two quadratic critical points $-1, 1$ and two critical values $-2, 2$. Each of the two circles of radii $R > 1$ and $1/R$ centered at 0 are mapped by f homeomorphically onto the ellipse E_R centered at 0 and having major axis $R(1 + R^{-1})$ on the real axis and minor axis $R(1 - R^{-1})$ on the imaginary axis. The restriction of f to the annulus $C_R = \{z \in \mathbb{C} : R^{-1} < |z| < R\}$ is a degree-2 covering map of $E_R \setminus \{-2, 2\}$. By the Schwarz lemma, the hyperbolic distance between the points -2 and 2 in the hyperbolic metric of E_R is a strictly monotone function of R, which tends to 0 as $R \to \infty$ and tends to ∞ as $R \to 1$. Since it is clearly continuous, it follows that this mapping is onto $\mathbb{R}^+$. Therefore, given any two points $p, q \in \mathbb{D}$ a unique value of $R > 1$ exists such that the hyperbolic distance between -2 and 2 in the hyperbolic metric of E_R is equal to the hyperbolic distance between p and q. Hence there exists a holomorphic diffeomorphism from E_R onto $\mathbb{D}$ that maps -2 into p and 2 into q. The composition of the restriction of f to C_R with this diffeomorphism gives the required covering map $\phi_{p,q}$, provided that we take $m(d) = \mod C_R = \pi/\log R$. $\qquad \square$

Theorem 4.3.7 *If* $\phi \colon \mathbb{D} \to \mathbb{D}$ *is a* K-*quasiconformal homeomorphism then*

$$\frac{1}{K} m(d(p, q)) \leq m(d(\phi(p), \phi(q))) \leq K m(d(p, q)) ,$$

where d *is the hyperbolic distance. In particular, the family of* K-*quasiconformal homeomorphisms of the disk is equicontinuous with respect to the hyperbolic distance.*

Proof Given p, q we can lift ϕ to a K-quasiconformal homeomorphism $\hat{\phi}: A \to A'$ where A is the annulus covering $\mathbb{D} \setminus \{p, q\}$ and A' is the annulus covering $\mathbb{D} \setminus \{\phi(p), \phi(q)\}$, as given by the above lemma. $\qquad\square$

Corollary 4.3.8 (Compactness) *Given $K > 1$ and $R > 0$, the set of K-quasiconformal diffeomorphisms of the disk such that the hyperbolic distance between 0 and its image is less than or equal to R is a compact subset of the space of continuous complex-valued maps of the disk, endowed with the topology of uniform convergence on compact subsets.*

Proof Let $\phi_n: \mathbb{D} \to \mathbb{D}$ be a sequence of K-quasiconformal homeomorphisms such that $d(0, \phi_n(0)) \leq R$. Let R' be such that $m(R') < K^{-1} m(R)$. We claim that $d(\phi_n^{-1}(0), 0) \leq R'$. Indeed, we have

$$m(d(\phi_n^{-1}(0), 0)) \geq \frac{1}{K} m(d(0, \phi(0))) \geq \frac{1}{K} m(R) \geq m(R') \ .$$

The claim follows because m is monotone. Thus, passing to a subsequence if necessary, we may assume that the sequences $\phi_n(0)$ and $\phi_n^{-1}(0)$ both converge. Since the hyperbolic metric and the Euclidean metric are equivalent in any given hyperbolic ball centered at 0, the restrictions of both sequences to each hyperbolic ball are equicontinuous in the Euclidean metric. Therefore, by the Arzelá–Ascoli theorem we can, passing to a subsequence, assume that $\phi_n \to \phi$ and $\phi_n^{-1} \to \psi$ uniformly on compact subsets of $\mathbb{D}$. Since the hyperbolic distance from $\phi_n(0)$ to 0 is uniformly bounded, it follows from the inequality in theorem 4.3.7 that for each x the hyperbolic distance from $\phi_n(x)$ to 0 is also uniformly bounded. Hence $\phi(x)$ belongs to $\mathbb{D}$. Similarly, $\psi(x) \in \mathbb{D}$. From the convergence, it follows that ψ is the inverse of ϕ. Let us now prove that ϕ is K-quasiconformal. Let A be an annulus compactly contained in $\mathbb{D}$. For $\epsilon > 0$ small, choose an annulus A_ϵ compactly contained in A such that A_ϵ separates the two components of the complement of $\widehat{\mathbb{C}} \setminus A$ and $\operatorname{mod} A > \operatorname{mod} A - \epsilon$. Since ϕ_n converges uniformly to ϕ in the closure of A_ϵ, we have that $\phi_n(A_\epsilon)$ is contained in $\phi(A)$ for large enough n. Hence, by the monotonicity of the modulus, we have

$$\operatorname{mod} \phi(A) \geq \operatorname{mod} \phi_n(A_\epsilon) \geq \frac{1}{K} \operatorname{mod} A_\epsilon \geq \frac{1}{K} (\operatorname{mod} A - \epsilon) \ .$$

Therefore $\operatorname{mod} \phi(A) \geq K^{-1} \operatorname{mod} A$. The other inequality is proved in the same way using ϕ^{-1}. $\qquad\square$

Corollary 4.3.9 (Hölder continuity) *Let $K > 1$ and $0 < r < 1$. There exists a constant $C(R) > 0$ such that, if $\phi \colon \mathbb{D} \to \mathbb{D}$ is a K-quasiconformal homeomorphism with $|\phi(0)| < r$, then we have*

$$|\phi(z) - \phi(w)| \leq C(R)|z - w|^{1/K}$$

for all z, w with $|z|, |w| < R < 1$. □

Corollary 4.3.10 (Boundary value) *If $\psi \colon \mathbb{D} \to \mathbb{D}$ is a quasiconformal homeomorphism then ϕ extends to a homeomorphism from the closure of $\mathbb{D}$ onto the closure of $\mathbb{D}$.* □

Example 4.3.11 Consider the map $f_0(z) = z^2$, whose Julia set is the circle $\partial \mathbb{D}$. Consider the annulus $A_0 = \{z \in \mathbb{C}; r^{-1} < |z| < r\}$ and its pre-image $A_1 = f_0^{-1}(A_0)$. Then A_1 is compactly contained in A_0 and the restriction of f_0 to A_1 is a degree-2 covering map. If c is a complex number close to zero and $f_c(z) = z^2 + c$ then $A_1^c = f_c^{-1}(A_0)$ is also an annulus compactly contained in A_0 and $\cap_{n=0}^{\infty} f_c^{-n}(A_0) = J(f_c)$ is the Julia set of f_c. We are going to construct, using the so-called pull-back argument of Sullivan, a quasiconformal homeomorphism h of A_0 whose restriction to A_1 conjugates f_0 with f_c. In particular, h will map the circle $J(f_0)$ onto $J(f_c)$ if $|c|$ is small enough. We start by constructing a C^∞ diffeomorphism h_0 as follows. We set h_0 to be the identity in a small neighborhood of the boundary of A_0, disjoint from the boundary of A_1. Next we define h_0 in the pre-image of this neighborhood, which is a neighborhood of the boundary of A_1, by lifting the identity, i.e. by forcing the conjugacy relation. Since this is close to the identity, we can use a partition of unity to construct h_0 to be a C^∞ diffeomorphism which is the identity except in a small neighborhood of the boundary of A_1, where it conjugates f_0 with f_c. Being a C^∞ diffeomorphism, h_0 is K-quasiconformal for some K. Since f_0 restricted to A_1 and f_c restricted to A_1^c are holomorphic covering maps, there is a unique diffeomorphism $h_1 \colon A_1 \to A_1^c$ that is a lift of h_0 and coincides with h_0 in the boundary of A_1. We can extend h_1 to A_0 by setting it equal to h_0 in $A_0 \setminus A_1$. This new diffeomorphism is K-quasiconformal for the same K value and conjugates the two maps on $A_1 \setminus A_2$, where $A_n = f^{-n}(A_0)$. By pulling back again, we construct a diffeomorphism h_2 that, restricted to A_1, is a lift of h_1. By induction we construct a sequence h_n of K-quasiconformal homeomorphisms such that the restriction of h_{n+1} to A_1 is a lift of h_n, $h_{n+1} = h_n$ in $A_0 \setminus A_{n+1}$ and conjugates the two maps in $A_0 \setminus A_{n+1}$. Using corollary 4.3.8 we can deduce that h_n converges uniformly to a K-quasiconformal homeomorphism h that conjugates the two maps in $A_0 \setminus \cap A_n$ and hence, by continuity, on A_0 since the Julia set has empty interior. We observe now that the limiting homeomorphism h may be very wild. Notice that 1 is a repelling fixed point of f_0 and therefore $h(1)$ is a fixed point of f_c. Now if we choose c so that the derivative of f_c at this fixed point has non-zero imaginary part then, by the invariance of the Julia set, it follows that it has to spiral around this fixed point. Again, by invariance, this spiral behavior also occurs in the whole backward orbit of the fixed point, which is dense in the Julia set. So the Julia set of f_c is quasiconformally homeomorphic to a circle, but is very wild. Using the fact that f_c is expanding in a neighborhood of the Julia set for $|c|$ small, and the corresponding uniform control on the distortion of iterates, one can prove that in fact, for $c \neq 0$ but small, the Julia set has Hausdorff dimension greater than 1. This shows that a quasiconformal homeomorphism may have a wild behavior.

Next, we are going to discuss a certain analytic property of homeomorphisms between plane domains that is equivalent to quasiconformality. Recall that a continuous real function $\alpha \colon \mathbb{R} \to \mathbb{R}$ is *absolutely continuous* if it has a derivative at Lebesgue-almost-every point, its derivative is integrable and the fundamental theorem of calculus is satisfied: $\alpha(b) - \alpha(a) = \int_a^b \alpha'(t)\, dt$.

Definition 4.3.12 *A continuous function $\phi \colon U \subset \mathbb{C} \to \mathbb{C}$ is absolutely continuous on lines if its real part and its imaginary part are absolutely continuous on Lebesgue-almost-all horizontals and on Lebesgue-almost-all verticals.*

Definition 4.3.13 *We say that a continuous function $\phi \colon U \to \mathbb{C}$ belongs to the Sobolev class $\mathcal{H}^1$ if there exist locally integrable measurable functions $\partial \phi$ and $\bar{\partial}\phi$ such that*

$$\iint \partial \phi\, h\, dxdy = - \iint \phi\, \partial h\, dxdy$$

and

$$\iint \bar{\partial}\phi\, h\, dxdy = - \iint \phi\, \bar{\partial} h\, dxdy$$

for all C^∞ functions h with compact support.

A complete proof of the following theorem can be found in [A1].

Theorem 4.3.14 *Let $K > 1$. The statements below are equivalent.*

(i) $\phi\colon U \to V$ *is a K-quasiconformal homeomorphism.*

(ii) $\phi\colon U \to V$ *is an orientation-preserving homeomorphism and is absolutely continuous on lines, the measurable function $\mu = \bar{\partial}\phi/\partial\phi$ belongs to $L^\infty(U)$ and $\|\mu\|_\infty \leq (K-1)/(K+1)$.*

(iii) $\phi\colon U \to V$ *is an orientation-preserving homeomorphism and belongs to the Sobolev class $\mathcal{H}^1$, the measurable function $\mu = \bar{\partial}\phi/\partial\phi$ belongs to $L^\infty(U)$ and $\|\mu\|_\infty \leq (K-1)/(K+1)$.*

A mapping having the regularity of the above theorem has a derivative almost everywhere with respect to Lebesgue measure, and such a derivative maps an ellipse of eccentricity at most K onto a circle. Hence a quasiconformal map defines a measurable field of ellipses with essentially bounded eccentricity. This field of ellipses is mapped into a field of circles by the derivatives. The theorem that we will discuss in the next section states that, conversely, any measurable field of ellipses with bounded eccentricity is associated with a quasiconformal map in this way.

Remark 4.3.15 *To verify that a given homeomorphism is quasiconformal, it is not sufficient to check the existence of derivatives almost everywhere and to control the eccentricity of the corresponding field of ellipses. It is also necessary to check the regularity condition. Indeed, let $\alpha\colon [0,1] \to [0,1]$ be a Cantor function, that is, a function that is constant in each component of the "middle thirds" Cantor set, is continuous, non-decreasing and onto $[0,1]$ (such a function is also called a devil's staircase). Extend α to the whole real line, putting $\alpha(x) = 0$ for $x \leq 0$ and $\alpha(x) = 1$ for $x \geq 1$. The map $\mathbb{R}^2 \to \mathbb{R}^2$ defined by $\phi(x,y) = (x, y + \alpha(x))$ is a homeomorphism and has a derivative almost everywhere equal to the identity, but it is not quasiconformal since it is not conformal.*

From the above theorem, we obtain immediately another proof that K-quasiconformality is a local property.

Proposition 4.3.16 *Let $\phi\colon \mathbb{D} \to \mathbb{D}$ be a K-quasiconformal homeomorphism. Then ϕ is the restriction of a K-quasiconformal homeomorphism of the Riemann sphere.*

Proof We extend ϕ to a homeomorphism of the whole Riemann sphere using geometric inversion with respect to the boundary of $\mathbb{D}$, i.e. $\phi(z) = I \circ \phi \circ I(z)$. Clearly ϕ is a homeomorphism and K-quasiconformal in the complement of the boundary of $\mathbb{D}$. Since every vertical and every horizontal intersects the boundary of $\mathbb{D}$ in at most two points, the extended homeomorphism is absolutely continuous on lines. Hence it is K-quasiconformal. □

It is possible to give another infinitesimal characterization of the quasiconformality of homeomorphisms between domains in the plane, one which involves no regularity condition and no derivatives, only the metric. This goes as follows. Let $f: U \to V$ be a homeomorphism. For each $x \in U$ let

$$H_f(x) = \limsup_{r \to 0} \frac{\max\{|f(y) - f(x)| : |x - y| = r\}}{\min\{|f(y) - f(x)| : |x - y| = r\}} .$$

Then one can prove that f is quasiconformal if and only if there exist $H > 0$ such that $H(x) \leq H$ for all $x \in U$. This statement was greatly improved by a recent result of Heinonen and Koskela [HK], which is the following.

Theorem 4.3.17 *An orientation-preserving homeomorphism $f: U \to V$ that satisfies*

$$\liminf_{r \to 0} \frac{\max\{|f(y) - f(x)| : |x - y| = r\}}{\min\{|f(y) - f(x)| : |x - y| = r\}} \leq H ,$$

for some $H < \infty$ and for all $x \in U$, is quasiconformal.

Another important classical property of quasiconformal homeomorphisms is absolute continuity: sets of zero Lebesgue measure are preserved. A more quantitative version of this fact was obtained by Astala [As] in the following result.

Theorem 4.3.18 *Given $K > 1$, there exists $M_K > 0$ such that if $f: \mathbb{C} \to \mathbb{C}$ is a K-quasiconformal map that fixes $0, 1, \infty$ then*

$$|f(E)| \leq M_K |E|^{1/K} .$$

Astala's proof of the above theorem involves some dynamical ideas related to the Ruelle–Bowen thermodynamic formalism (see Chapter 3) and also the theory of holomorphic motions, which we will discuss in Chapter 5. As a consequence of this area distortion theorem, Astala also obtained a distortion result for the Hausdorff dimension of compact sets under quasiconformal maps.

Corollary 4.3.19 *Let* $f: U \to V$ *be a* K-quasiconformal diffeomorphism. If $C \subset U$ is a compact subset and $\dim_H C$ is its Hausdorff dimension, then*

$$\frac{1}{K}\left(\frac{1}{\dim_H C} - \frac{1}{2}\right) \leq \frac{1}{\dim_H f(C)} - \frac{1}{2} \leq K\left(\frac{1}{\dim_H C} - \frac{1}{2}\right).$$

Again, the proof can be found in [As].

4.4 The Ahlfors–Bers theorem

One of the most important results in the differential geometry of surfaces is the Gauss theorem on the existence of *isothermal coordinates*. These are local diffeomorphisms of open subsets, of a surface endowed with a smooth Riemannian metric, onto open subsets of the complex plane whose derivative, at each point, maps the set of unit vectors with respect to the metric onto a round circle in the plane. Hence this diffeomorphism is an isometry between the metric and a Riemannian metric in the image that is conformal with respect to the Euclidean metric. In coordinates, the isothermal parameters of Gauss are diffeomorphic solutions of a partial differential equation known as the Beltrami equation, which we will discuss below. In our language, the Gauss theorem states that any smooth function $\mu: U \to \mathbb{D}$ is locally the Beltrami coefficient of a quasiconformal diffeomorphism. The so-called *measurable Riemann mapping theorem* is a generalization of the Gauss theorem to measurable functions with L^∞ norm strictly less than 1. This theorem was proved by Morrey, who in 1938 established the existence of quasiconformal solutions. The holomorphic dependence on parameters was obtained by Ahlfors and Bers in 1961, by a method based on earlier work by Bojarskii, and culminated in the following theorem.

Theorem 4.4.1 (Ahlfors–Bers) *Let* U *be a domain of the Riemann sphere.*

(i) *Given a measurable function* $\mu: U \to \mathbb{D}$ *such that* $\|\mu\|_\infty < 1$, *there exists a quasiconformal homeomorphism* $f: U \to V$ *that is a solution of the Beltrami equation*

$$\bar{\partial} f = \mu \, \partial f.$$

Two such solutions differ by post-composition with a holomorphic diffeomorphism. In particular, if U is the whole Riemann sphere then there exists a unique homeomorphic solution that fixes three given points.

(ii) *Let Λ be an open subset of some complex Banach space and consider a map $\Lambda \times \mathbb{C} \to \mathbb{D}$, $(\lambda, z) \mapsto \mu_\lambda(z)$, satisfying the following properties.*

    (a) *For every λ the mapping $\mathbb{C} \to \mathbb{D}$ given by $z \mapsto \mu_\lambda(z)$ is measurable, and $\|\mu_\lambda\|_\infty \leq k$ for some fixed $k < 1$.*

    (b) *For Lebesgue-almost-all z, the mapping $\Lambda \to \mathbb{D}$ given by $\lambda \mapsto \mu_\lambda(z)$ is holomorphic.*

For each λ, let f_λ be the unique quasiconformal homeomorphism of the Riemann sphere that fixes $0, 1, \infty$ and whose Beltrami coefficient is μ_λ. Then the mapping $\lambda \mapsto f_\lambda(z)$ is holomorphic for all z.

We shall present a proof only of part (i) of the Ahlfors–Bers theorem, which is in fact Morrey's theorem. The proof will be quite long, involving several interesting ideas. In order to facilitate understanding, we divide the proof into steps.

4.4.1 Proof of the Ahlfors–Bers theorem

We want to solve the Beltrami equation on a given domain $U \subseteq \widehat{\mathbb{C}}$. Here we search for a quasiconformal homeomorphism $f : U \to f(U) \subseteq \widehat{\mathbb{C}}$ whose complex dilatation μ_f agrees with a given $\mu \in L^\infty(U)$ (with $\|\mu\|_\infty < 1$) at almost every point of U. Extending μ to be zero on $\widehat{\mathbb{C}} \setminus U$, we may assume that $U = \widehat{\mathbb{C}}$. There are various ways in which the solutions can be *normalized*. For instance, we could require that the points $0, 1, \infty \in \widehat{\mathbb{C}}$ remain fixed. Alternatively, we could require that the solution be tangent to the identity at ∞. We will use this second normalization in the discussion below. Whichever normalization one chooses, the normalized solution of the Beltrami equation for a given Beltrami differential on the Riemann sphere will be *unique*.

Step 1. It suffices to consider the case when the support of μ is compact. Indeed, suppose that we can solve the Beltrami equation for any Beltrami form with support in the unit disk. Given an arbitrary $\mu \in L^\infty(\widehat{\mathbb{C}})$ with $\|\mu\|_\infty < 1$, define $\mu_0 \in L^\infty(\widehat{\mathbb{C}})$ by

$$\mu_0(z) = \begin{cases} \mu(z) & \text{if } z \in \widehat{\mathbb{C}} \setminus \mathbb{D}, \\ 0 & \text{if } z \in \mathbb{D}. \end{cases}$$

There exists a normalized solution $f_0 : \widehat{\mathbb{C}} \to \widehat{\mathbb{C}}$ to the Beltrami equation $\overline{\partial} f = \mu_0 \, \partial f$. To see why, take $\mu_0^*(z) = \mu_0(1/z) z^2 / \overline{z}^2$, note that the support of μ_0^* is now contained in $\mathbb{D}$ and let $g : \widehat{\mathbb{C}} \to \widehat{\mathbb{C}}$ be the corresponding normalized solution to $\overline{\partial} f = \mu_0^* \partial f$, which exists by assumption; then define $f_0(z) = 1/g(1/z)$. Next, let $W = f_0(\mathbb{D})$ and note that W is bounded. For all $w \in W$, define

$$\mu_1(w) = \mu(f_0^{-1}(w)) \, f_0'(f_0^{-1}(w)) \big/ \, \overline{f_0'(f_0^{-1}(w))} \, .$$

This makes sense because f_0 is conformal in $\mathbb{D}$ (so that $f_0'(z)$ exists for all $z \in \mathbb{D}$, and *a priori* it can only vanish at countably many points). Set $\mu_1(w) = 0$ for all $w \notin \mathbb{D}$. Then $\mu_1 \in L^{\infty}(\widehat{\mathbb{C}})$, we have $\|\mu_1\|_{\infty} < 1$ and μ_1 has compact support (contained in W). Let $f_1 : \widehat{\mathbb{C}} \to \widehat{\mathbb{C}}$ be the normalized solution to $\overline{\partial} f = \mu_1 \partial f$, which once again exists by assumption. Taking $f : \widehat{\mathbb{C}} \to \widehat{\mathbb{C}}$ to be the composition $f = f_1 \circ f_0$, we get a normalized quasiconformal map, and the formula for the complex dilatation of composite maps shows at once that $\mu_f = \mu$.

Step 2. Given the previous step, from now on we restrict our attention to Beltrami forms μ having compact support in $\mathbb{C}$. Let us examine more closely those quasiconformal homeomorphisms whose complex dilatations are of this type.

Lemma 4.4.2 *Let* $f : \widehat{\mathbb{C}} \to \widehat{\mathbb{C}}$ *be a quasiconformal homeomorphism with the following properties:*

(i) *f is tangent to the identity at ∞, in the sense that $f(z) = z + \mathcal{O}(1/z)$ for $|z| \to \infty$;*

(ii) *$\mu_f \in L^{\infty}(\widehat{\mathbb{C}})$ has compact support in $\mathbb{C}$;*

(iii) *∂f is locally in L^p for some $p > 2$.*

Then, for all $z \in \mathbb{C}$, we have

$$f(z) = z + \frac{1}{2\pi i} \iint_{\mathbb{C}} \frac{\mu_f(\zeta) \partial f(\zeta)}{\zeta - z} \, d\zeta \wedge d\overline{\zeta} \, . \tag{4.2}$$

Proof Note that the hypotheses imply that $\overline{\partial} f$ is locally in L^p also. Given $z \in \mathbb{C}$, let $D = D(0, R)$ be a large disk containing z and the support of μ_f. Then, by Pompeiu's formula (proposition 2.1.3), we have

$$f(z) = \frac{1}{2\pi i} \int_{\partial D} \frac{f(\zeta)}{\zeta - z} \, d\zeta + \frac{1}{2\pi i} \iint_{D} \frac{\overline{\partial} f(\zeta)}{\zeta - z} \, d\zeta \wedge d\overline{\zeta} \, . \tag{4.3}$$

Now, on the one hand we have $\overline{\partial} f(\zeta) = \mu_f(\zeta) \partial f(\zeta)$ for almost all ζ. On the other hand, for all $\zeta \in \partial D$ we can write $f(\zeta) = \zeta + \psi(\zeta)$, where ψ is holomorphic on $\widehat{\mathbb{C}} \setminus D$ and $\psi(\zeta) = \mathcal{O}(1/\zeta)$; this gives us

$$\frac{1}{2\pi i} \int_{\partial D} \frac{f(\zeta)}{\zeta - z} \, d\zeta \;=\; z + \frac{1}{2\pi i} \int_{\partial D} \frac{\psi(\zeta)}{\zeta - z} \, d\zeta \,.$$

This last integral is independent of R (as long as D contains z and the support of μ_f), and an easy estimate shows that its value is $\mathcal{O}(1/R)$. Hence it is necessarily equal to zero. Using these facts in (4.3) we get (4.2) as desired. $\qquad\square$

Step 3. Following [CG], we shall denote by $\mathrm{QC}(k,R)$ the set of all quasiconformal homeomorphisms $f : \widehat{\mathbb{C}} \to \widehat{\mathbb{C}}$ such that $\|\mu_f\|_\infty \le k < 1$ and which satisfy (i)–(iii) of lemma 4.4.2. One can show with the help of Koebe's one-quarter theorem that if $f \in \mathrm{QC}(k,R)$ then $f^{-1} \in \mathrm{QC}(k,4R)$. Now we have the following Hölder estimate for maps in $\mathrm{QC}(k,R)$.

Lemma 4.4.3 *If $f \in \mathrm{QC}(k,R)$ and $p > 2$ is such that $\partial f \in L^p_{\mathrm{loc}}$ then for all $z_1, z_2 \in D(0,R)$ we have*

$$|f(z_1) - f(z_2)| \;\le\; C\,|z_1 - z_2|^{1-2/p} \,,$$

where $C > 0$ is a constant depending only on k and R and on the L^p norm of ∂f in $D(0,R)$.

Proof Given $z_1, z_2 \in D(0,R)$, we calculate $f(z_1) - f(z_2)$ by means of formula (4.2) and get

$$f(z_1) - f(z_2) \;=\; z_1 - z_2 + \frac{z_1 - z_2}{2\pi i} \iint_{\mathbb{C}} \frac{\mu_f(\zeta)\partial f(\zeta)}{(\zeta - z_1)(\zeta - z_2)} \, d\zeta \wedge d\overline{\zeta} \,.$$

Applying Hölder's inequality to the integral on the right-hand side, we obtain

$$|f(z_1) - f(z_2)| \;=\; |z_1 - z_2| \left(1 + \frac{k\,\|\partial f\|_p}{2\pi} \left[\iint_{\mathbb{C}} \frac{|d\zeta \wedge d\overline{\zeta}|}{(|\zeta - z_1|\,|\zeta - z_2|)^q} \right]^{1/q} \right). \tag{4.4}$$

In order to estimate this last integral, let $2\rho = |z_1 - z_2|$ and define $D_1 = \{\zeta : |\zeta - z_1| < \rho\}$ and $D_2 = \{\zeta : |\zeta - z_2| < \rho\}$. Note that, since D_1 and D_2 are disjoint, we have

$$\iint_{D_1} \frac{|d\zeta \wedge d\overline{\zeta}|}{(|\zeta - z_1|\,|\zeta - z_2|)^q} \;\le\; \frac{1}{\rho^q} \iint_{D_1} \frac{|d\zeta \wedge d\overline{\zeta}|}{|\zeta - z_1|^q} \,.$$

We will compute the integral on the right-hand side using polar coordinates, so that

$$\iint_{D_1} \frac{|d\zeta \wedge d\bar\zeta|}{|\zeta - z_1|^q} = |-2i| \int_0^\rho \int_0^{2\pi} \frac{r \, dr d\theta}{r^q} = \frac{4\pi}{2-q} \rho^{2-q} .$$

Therefore

$$\iint_{D_1} \frac{|d\zeta \wedge d\bar\zeta|}{(|\zeta - z_1||\zeta - z_2|)^q} \leq \frac{2^{2q}\pi}{2-q} |z_1 - z_2|^{2-2q} .$$

Similarly, we have

$$\iint_{D_2} \frac{|d\zeta \wedge d\bar\zeta|}{(|\zeta - z_1||\zeta - z_2|)^q} \leq \frac{2^{2q}\pi}{2-q} |z_1 - z_2|^{2-2q} .$$

Now using the symmetry about the line $|\zeta - z_1| = |\zeta - z_2|$, we get

$$\iint_{\mathbb{C}\backslash(D_1 \cup D_2)} \frac{|d\zeta \wedge d\bar\zeta|}{(|\zeta - z_1||\zeta - z_2|)^q} \leq 2 \iint_{\mathbb{C}\backslash D_1} \frac{|d\zeta \wedge d\bar\zeta|}{|\zeta - z_1|^{2q}}$$
$$= 4 \int_\rho^\infty \int_0^{2\pi} \frac{r \, dr d\theta}{r^{2q}}$$
$$= \frac{2^{2q}\pi}{q-1} |z_1 - z_2|^{2-2q} .$$

Putting all these estimates together back into (4.4) and taking into account that $|z_1 - z_2| \leq (2R)^{2/p}|z_1 - z_2|^{1-2/p}$, we deduce after some simple computations that

$$|f(z_1) - f(z_2)| \leq C |z_1 - z_2|^{2/q-1} = C |z_1 - z_2|^{1-2/p}$$

for a certain constant $C = C(k, R, \|\partial f\|_p) > 0$, as was to be proved. $\square$

Step 4. Note that if $f \in \mathrm{QC}(k, R)$ then $\psi : \widehat{\mathbb{C}} \to \widehat{\mathbb{C}}$ given by $\psi(z) = f(z) - z$ has the following properties (where $\mu = \mu_f$):

(1) $\partial\psi \in L^p(\mathbb{C})$ for some $p > 2$;
(2) ψ is holomorphic near ∞ and $\psi(z) = \mathcal{O}(1/z)$ as $|z| \to \infty$;
(3) $\bar\partial\psi = \mu + \mu\partial\psi$ almost everywhere.

Conversely, if ψ satisfies these conditions and $\|\mu\|_\infty = k < 1$ then it is reasonable to expect that $f(z) = z + \psi(z)$ will be a quasiconformal homeomorphism belonging to $\mathrm{QC}(k, R)$. This indicates that we should consider the *generalized Beltrami equation* (GBE) given by

$$\bar\partial\psi = \nu + \mu\,\partial\psi, \tag{4.5}$$

where μ and ν are given L^∞ functions.

Step 5. In order to solve the GBE, we need a crucial estimate concerning a certain singular integral operator known as the *Beurling transform*. First note that if ψ is a solution of (4.5), say for μ, ν with compact support, and $\partial\psi, \overline{\partial}\psi \in L^p(\widehat{\mathbb{C}})$ then

$$\psi(z) = \frac{1}{2\pi i} \iint_{\mathbb{C}} \frac{\overline{\partial}\psi(\zeta)}{\zeta - z} \, d\zeta \wedge d\overline{\zeta} \,. \tag{4.6}$$

In other words, we have $\psi = T(\overline{\partial}\psi)$, where T is the singular integral operator given by

$$T\varphi(z) = \frac{1}{2\pi i} \iint_{\mathbb{C}} \frac{\varphi(\zeta)}{\zeta - z} \, d\zeta \wedge d\overline{\zeta} \,.$$

This is an unbounded operator, which is difficult to deal with directly. But we can write, formally at least,

$$\partial\psi = \partial\left(T(\overline{\partial}\psi)\right) = S(\overline{\partial}\psi) \,,$$

where S is the *Beurling transform*, given by

$$S\varphi(z) = \frac{1}{2\pi i} \iint_{\mathbb{C}} \frac{\varphi(\zeta)}{(\zeta - z)^2} \, d\zeta \wedge d\overline{\zeta} \,. \tag{4.7}$$

This expression should be understood as a Cauchy principal value. In other words, if $\varphi \in C_0^\infty(\mathbb{C})$, say, we define

$$S\varphi(z) = \lim_{\varepsilon \to 0} \frac{1}{2\pi i} \iint_{|\zeta - z| > \varepsilon} \frac{\varphi(\zeta)}{(\zeta - z)^2} \, d\zeta \wedge d\overline{\zeta} \,.$$

A straightforward computation shows that $\|S\varphi\|_2 = \|\varphi\|_2$ for all $\varphi \in C_0^\infty(\mathbb{C})$ (exercise 4.6). Since $C_0^\infty(\mathbb{C})$ is dense in $L^2(\mathbb{C})$, it follows that S extends to an isometry of $L^2(\mathbb{C})$. Knowing that S is a bounded linear operator in L^2 is unfortunately not enough for our purposes, because we want to solve the GBE in L^p with $p > 2$. The really crucial point for us is that S extends to a bounded operator in L^p for every $1 < p < \infty$! This is the essence of the following difficult result.

Theorem 4.4.4 (Calderón–Zygmund) *The Beurling transform extends to a bounded linear operator $S : L^p(\mathbb{C}) \to L^p(\mathbb{C})$ for all $1 < p < \infty$. More precisely, there exists a continuous logarithmically convex function $p \mapsto C_p > 0$ with $C_2 = 1$ such that $\|S\varphi\|_p \leq C_p \|\varphi\|_p$ for all $\varphi \in L^p(\mathbb{C})$.*

For a complete proof of this theorem, see [A1, pp. 106–115]. The Calderón–Zygmund inequality allows us to solve the GBE in a fairly easy way.

Theorem 4.4.5 (Generalized Beltrami) *Let $\mu, \nu \in L^\infty(\mathbb{C})$ have compact support, and suppose that $\|\mu\|_\infty < 1$. Then there exists a unique continuous map $\psi : \widehat{\mathbb{C}} \to \widehat{\mathbb{C}}$ which is holomorphic near ∞, satisfies $\psi(z) = \mathcal{O}(1/z)$ and is such that*

$$\overline{\partial}\psi(z) = \mu(z)\,\partial\psi(z) + \nu(z) \tag{4.8}$$

for almost every $z \in \widehat{\mathbb{C}}$. Moreover $\partial\psi, \overline{\partial}\psi \in L^p(\mathbb{C})$ for some $p > 2$.

Proof Let $k = \|\mu\|_\infty < 1$ and let us fix $p > 2$ such that $kC_p < 1$, where C_p is the constant of theorem 4.4.4. If a solution ψ exists, $\overline{\partial}\psi$ must be a solution in L^p of the functional linear equation

$$\phi = \nu + \mu S(\phi) . \tag{4.9}$$

We are now in very familiar territory. We solve this equation taking, say, $\varphi_0 = \mu S(\nu)$ and defining inductively $\varphi_{n+1} = \mu S(\varphi_n) \in L^p(\mathbb{C})$. Then the *Neumann series*

$$\phi = \nu + \sum_{n \geq 0} \varphi_n$$

converges geometrically in $L^p(\mathbb{C})$, because for all $n \geq 0$ we have

$$\|\varphi_{n+1}\|_p \leq \|\mu\|_\infty \|S(\varphi_n)\|_p \leq kC_p\|\varphi_n\|_p ,$$

by the Calderón–Zygmund inequality. Moreover

$$\begin{aligned}
\mu S(\phi) &= \mu S(\nu) + \sum_{n \geq 0} \mu S(\varphi_n) \\
&= \mu S(\nu) - \varphi_0 + \sum_{n \geq 0} \varphi_n \\
&= \phi - \nu ,
\end{aligned}$$

whence we deduce that ϕ is the unique solution to (4.9) as claimed. Note that ϕ has compact support (as does each φ_n). To get ψ from ϕ, we have to solve the $\overline{\partial}$-problem $\overline{\partial}\psi = \phi$. The idea now is that we can simply use (4.6). We may write

$$\psi(z) = \frac{1}{2\pi i} \iint_{\mathbb{C}} \frac{\phi(\zeta)}{\zeta - z}\, d\zeta \wedge d\overline{\zeta} ,$$

which is certainly well defined since $\phi \in L^p(\mathbb{C})$. A detailed verification that ψ is the desired solution is left to the reader (see exercise 4.7). It is clear from this formula that $|\psi(z)| = \mathcal{O}(1/|z|)$ as $|z| \to \infty$. Note also that $\partial \psi, \overline{\partial} \psi \in L^p(\mathbb{C})$. Finally, the same argument as that used in the proof of lemma 4.4.3 shows that ψ is Hölder continuous (with exponent $1 - 2/p$) and therefore continuous (see exercise 4.8). □

Observe that the above proof also shows that $\phi = \overline{\partial} \psi$ has L^p norm bounded by $(1 - kC_p)^{-1} \|\nu\|_p$. This has the important consequence that the Hölder constant C for $f \in \mathrm{QC}(k, R)$ in lemma 4.4.3 depends only on k and R. Hence, by the Arzelá–Ascoli theorem, the family $\mathrm{QC}(k, R)$ is compact in the topology of uniform convergence on compact subsets of the complex plane. This remark will be very useful in the last step of the proof.

Step 6. Having solved the GBE in the previous step, we might grow over-confident and jump to the conclusion that the solution to the Beltrami problem is now complete: all we have to do is set $\nu = \mu$ in (4.5), solve for ψ using the method described in step 5 and then take $f(z) = z + \psi(z)$, right? Not quite! We don't want merely a solution to the Beltrami equation, we want the solution to be a topological mapping, that is to say a (quasiconformal) *homeomorphism*. Checking directly that the solution is a homeomorphism for arbitrary μ seems too difficult. Instead, we will first assume that μ is C^1, show that the corresponding f is a C^1 diffeomorphism and then handle the general case via an approximation argument. We need the following lemma.

Lemma 4.4.6 *Let $\Omega \subseteq \mathbb{C}$ be a simply connected domain, and fix some point $a \in \Omega$. Suppose that $u, v : \Omega \to \mathbb{C}$ are continuous, have locally integrable partial derivatives and satisfy $\overline{\partial} u = \partial v$ throughout Ω. Then $f : \Omega \to \mathbb{C}$ given by*

$$f(z) = \int_a^z u(\zeta) \, d\zeta + v(\zeta) \, d\overline{\zeta}$$

is a well-defined C^1 function, with $\partial f = u$ and $\overline{\partial} f = v$.

Proof If u and v are themselves C^1 then the condition $\overline{\partial} u = \partial v$ means that the 1-form $\omega = u \, d\zeta + v \, d\overline{\zeta}$ is closed in Ω, hence exact, and the desired result follows from Green's formula. To reduce the general case to this one, use L^1 smoothing sequences (see section 2.1) to approximate

u, v by $u_n, v_n \in C^1$ respectively, where u_n, v_n satisfy $\overline{\partial} u_n = \partial v_n$ for all n, and then apply the dominated convergence theorem as $n \to \infty$. $\qquad\square$

With this lemma in hand, we proceed as follows. Suppose that μ is C^1 (and has compact support). We already know from theorem 4.4.5 (for $\nu = \mu$) that there exists a continuous $f : \widehat{\mathbb{C}} \to \widehat{\mathbb{C}}$, with distributional partial derivatives locally in L^p (for some $p > 2$) such that $\overline{\partial} f = \mu \, \partial f$ almost everywhere. If f is a C^1 diffeomorphism then $u = \partial f$ and $v = \overline{\partial} f$ are continuous, the 1-form $u \, d\zeta + v \, d\overline{\zeta}$ is closed and thus $\overline{\partial} u = \partial v$, at least in the distributional sense. From this and the fact that $v = \mu u$ (the Beltrami equation) we deduce that u satisfies the equation

$$\overline{\partial} u = \mu \, \partial u + \mu_z u \,, \tag{4.10}$$

where $\mu_z = \partial \mu$ is continuous with compact support. In addition, since

$$0 \neq \operatorname{Jac} f = |\partial f|^2 - |\overline{\partial} f|^2 = \left(1 - |\mu|^2\right) |u|^2 \,,$$

we have $u(\zeta) \neq 0$ for all ζ. Therefore, we can find a continuous function σ such that $u = e^\sigma$. Substituting this information into (4.10), we see that σ satisfies

$$\overline{\partial}(e^\sigma) = \mu \, \partial(e^\sigma) + \mu_z e^\sigma$$

or, cancelling out the factor e^σ from both sides,

$$\overline{\partial} \sigma = \mu \, \partial \sigma + \mu_z \,. \tag{4.11}$$

But then we can simply invert the entire argument. Indeed, applying theorem 4.5 with $\nu = \mu_z$, we get a *continuous* function σ that satisfies (4.11) in the distributional sense. We can take $u = e^\sigma$, form $v = \mu e^\sigma$ and then invoke lemma 4.4.6: this is legitimate since both u and v are continuous and $\overline{\partial} u = \partial v$. We get a C^1 map $f : \widehat{\mathbb{C}} \to \widehat{\mathbb{C}}$ with $\operatorname{Jac} f \neq 0$ everywhere. Thus, f is a local C^1 diffeomorphism everywhere and, since $\widehat{\mathbb{C}}$ is simply connected, f must be a *global* C^1 diffeomorphism.

Step 7. Finally, if $\mu \in L^\infty(\mathbb{C})$ with $k = \|\mu\|_\infty < 1$ is an arbitrary Beltrami differential with compact support, we can approximate μ by a sequence $\mu_n \in L^\infty(\mathbb{C})$ such that $\|\mu_n - \mu\|_\infty \to 0$ and $\mu_n \in C^1$ for all n. If the support of μ is contained in the disk $D(0, R)$, we can choose our approximating sequence so that the support of each μ_n is contained in this disk also. Let f_{μ_n} be the solution of the Beltrami equation for μ_n obtained in step 6. Then for all n we have $f_{\mu_n} \in \operatorname{QC}(k', R)$ for some fixed $k \leq k' < 1$. We also have $f_{\mu_n}^{-1} \in \operatorname{QC}(k', 4R)$, by the remark at the beginning of step 3. Applying lemma 4.4.3 and the remark at the end of step 5, we deduce that both the families $\{f_{\mu_n}\}$ and $\{f_{\mu_n}^{-1}\}$ are equicontinuous and uniformly bounded on any compact subset of $\widehat{\mathbb{C}}$. Passing to a subsequence if necessary, we deduce that $\{f_{\mu_n}\}$ converges uniformly in the whole sphere to some $f \in QC(k', R)$. From these facts it follows at once that $\mu_f = \mu$, so that $\bar{\partial}f = \mu\partial f$, and therefore this quasiconformal map f is the desired solution to the Beltrami equation for μ.

This completes the proof of the first part of the Ahlfors–Bers theorem (Morrey's theorem). The proof of the second part is similar (but shorter), hence we will omit it. The reader may consult [A1, pp. 100–6] for full details. $\Box$

We close this section with a couple of remarks.

Remark 1   Recently, A. Douady [Dou] found a more elementary proof of the Ahlfors–Bers theorem that relies only on classical Fourier analysis on $L^2(\mathbb{C})$.

Remark 2   There is a striking generalization of the measurable Riemann mapping theorem for a certain class of homeomorphisms f for which $\|\mu_f\|_\infty = 1$, due to G. David [Da]. In order to state it, we need the following definition. We say that an orientation-preserving homeomorphism $f : U \to V$ between domains $U, V \subseteq \widehat{\mathbb{C}}$ is a *David homeomorphism* if

    (a) f is absolutely continuous on lines;

    (b) there exist constants $C > 0$, $\alpha > 0$ and $0 < \delta_0 < 1$ such that for all $0 < \delta < \delta_0$ we have

$$\text{Area}\{z \in U : |\mu_f(z)| > 1 - \delta \} \leq Ce^{-\alpha/\delta}.$$

Here "Area" means the *spherical* area. Clearly, every quasiconformal homeomorphism is a David homeomorphism. But general David homeomorphisms behave quite differently from quasiconformal homeomorphisms. For instance, the inverse of a David homeomorphism is not necessarily a David homeomorphism. Nevertheless, we have the following amazing result.

Theorem 4.4.7 (David) *If $\mu : U \to \mathbb{D}$ is a David–Beltrami form then μ can be integrated; in other words, there exists a David homeomorphism $f : U \to V \subseteq \widehat{\mathbb{C}}$ such that $\bar{\partial}f(\zeta) = \mu(\zeta)\,\partial f(\zeta)$ for almost every $\zeta \in U$. If $g : U \to W \subseteq \widehat{\mathbb{C}}$ is another David homeomorphism with the same property, then $g \circ f^{-1} : V \to W$ is conformal.*

This theorem has some nice dynamical applications. It has been used by P. Haissinsky [Ha] to study the dynamics of parabolic points of rational maps. More recently, C. Petersen and S. Zakeri [PZ] used David's theorem to show that for Lebesgue-almost-every $0 < \theta < 1$ the Julia set of the quadratic polynomial $P_\theta : z \mapsto e^{2\pi i\theta}z + z^2$ is locally connected and has area zero.

4.5 First dynamical applications

In order to illustrate the usefulness of the Ahlfors–Bers theorem in complex dynamics, let us discuss a few striking applications.

Theorem 4.5.1 *In the spaces of rational maps of degree d or polynomials of degree d, each quasiconformal conjugacy class of maps is connected.*

Proof Let f and g be two quasiconformally conjugate rational maps, and let h be a quasiconformal conjugacy between f and g. Its Beltrami coefficient μ is invariant under the dynamics of f, in the sense that the corresponding ellipse field is preserved by the derivative of f at Lebesgue-almost-every point. If w is a complex number in the closed unit disk, then the Beltrami coefficient $w\mu$ is also invariant under the dynamics of f. Let h_w be the unique quasiconformal homeomorphism with Beltrami coefficient $w\mu$ that coincides with h at the three points $0, 1, \infty$. By the Ahlfors–Bers theorem, this gives a continuous family of homeomorphisms connecting $h_1 = h$ to a Möbius transformation h_0. Furthermore, the map $g_w = h_w \circ f \circ h_w^{-1}$ is locally 1-quasiconformal and, therefore, a conformal mapping. Thus g_w, $w \in [0, 1]$, is a path of rational maps connecting g with the rational map $h_0 \circ f \circ h_0^{-1}$. Now, if we take a path of Möbius transformations connecting h_0 with the identity we get a path of rational mappings connecting $h_0 \circ f \circ h_0^{-1}$ with f. $\square$

Another simple application of the Ahlfors–Bers theorem is Sullivan's proof of the finiteness of the number of periodic cycles of the Fatou components.

Theorem 4.5.2 *If f is a rational map of degree d then the number of Herman rings of f is bounded by $2d - 1$.*

Proof Suppose, by contradiction, that the number of orbits of Herman rings for f is $N > 2d - 1$. We want to construct a holomorphic family f_w of rational maps, for w in a neighborhood of 0 in $\mathbb{C}^N$, such that $f_w \neq f_{w'}$ if $w \neq w'$, but this is not possible because N is greater than the dimension of the space of rational maps. So, to construct this family of maps we start by constructing a family μ_i, $i = 1, \ldots, N$, of Beltrami coefficients, each supported in the grand orbit of a Herman ring and invariant under f. To construct each Beltrami coefficient we consider a Herman ring of period k. Then the restriction of f^k to this ring is conformally equivalent to an irrational rotation of a round annulus. Considering a non-zero Beltrami coefficient invariant under such a rotation and taking the pull-back by the conformal equivalence, we get a Beltrami coefficient in the Herman ring that is invariant under f^k. Pulling back by iterates of f we get a Beltrami coefficient with the same L^∞ norm in the full grand orbit of that Herman ring. Now we consider the holomorphic family μ_w of Beltrami coefficients defined to be equal to 0 in the complement of the grand orbits of all Herman rings and equal to $\sum w_i \mu_i$ otherwise, where the w_i are complex numbers with $|w_i| \leq 1$. Let h_w be the holomorphic family of quasiconformal homeomorphisms with Beltrami coefficients μ_w given by the Alhfors–Bers theorem. Since each μ_w is invariant under f as a Beltrami coefficient, it follows that $f_w = h_w \circ f \circ h_w^{-1}$ is a family of rational maps. This is a holomorphic locally injective family of rational maps of degree d, which is not possible since N is greater than the dimension of the space of rational maps of degree d. $\qquad\square$

4.6 The no-wandering-domains theorem

We now turn our attention to a theorem that is a true landmark in the subject of complex dynamics. Historically, the first application of the measurable Riemann mapping theorem to the dynamics of rational maps was Sullivan's *no-wandering-domains theorem*. In proving this theorem here, we shall follow the extremely elegant approach of McMullen [McM4], which is based on the study of the infinitesimal deformations of a rational map. This approach is quite natural and avoids certain technical complications concerning the boundary behavior of conformal maps, which forced Sullivan to use Carathéodory's theory of *prime ends*; see [Su]. For a slightly different approach to Sullivan's theorem using *harmonic* Beltrami differentials, see [McS] or [MNTU].

Theorem 4.6.1 (Sullivan) *Let $f : \widehat{\mathbb{C}} \to \widehat{\mathbb{C}}$ be a rational map. Then each connected component of the Fatou set of f is eventually periodic.*

Recall that a connected component U of the Fatou set of f is a *wandering domain* if its forward images $f^n(U)$, $n \geq 0$, are pairwise disjoint. Sullivan's theorem is equivalent to the statement that there are no wandering domains for f, hence the name. Before going on to prove this statement, it is wise to take first the following reduction step due to I. N. Baker.

Lemma 4.6.2 *If f has a wandering domain then it has a simply connected wandering domain.*

Proof Let U be a wandering domain for f, and consider its forward images $U_n = f^n(U)$, $n \geq 0$. Since these are pairwise disjoint and f has only a finite number of critical points, we may assume – discarding the first few U_n if necessary – that no U_n contains a critical point of f. Thus each $f : U_n \to U_{n+1}$ is a covering map, and it is a local isometry if we endow each U_n with its hyperbolic metric. Suppose that $U = U_0$ is not simply connected. Note that no U_n can be a punctured disk, because the Julia set of f is perfect. Hence for each n we can find a closed geodesic in the hyperbolic metric of U_n, say $\gamma_n \subset U_n$, such that $f(\gamma_n) = \gamma_{n+1}$. Since f is a local isometry in each U_n, the hyperbolic length of γ_{n+1} in U_{n+1} is at most equal to the hyperbolic length of γ_n in U_n. We claim that $\mathrm{diam}_{\widehat{\mathbb{C}}}(\gamma_n) \to 0$ in the spherical metric as $n \to \infty$. This follows from the fact that the spherical diameter of the largest disk contained in U_n must go to zero as $n \to \infty$, because the U_n are pairwise disjoint, and from a simple comparison of the spherical and hyperbolic metrics in each U_n. Now, f is a Lipschitz map in the spherical metric. This means that for all sufficiently large n the small components of $\widehat{\mathbb{C}} \setminus \gamma_n$ (where "small" means having a spherical diameter comparable with the spherical diameter of γ_n) must be mapped into small components of $\widehat{\mathbb{C}} \setminus \gamma_{n+1}$. Hence the iterates of f restricted to such a small component form a normal family. But this is impossible, because each component of $\widehat{\mathbb{C}} \setminus \gamma_n$ meets the Julia set of f. This contradiction shows that U is simply connected as asserted. $\square$

In order to rule out the existence of wandering disks for f, we shall examine the infinitesimal deformations of f.

We recall that the space $\mathrm{Rat}_d(\widehat{\mathbb{C}})$ of rational maps of degree d can be identified with an open subset of $\mathbb{C}P^{2d+1}$ (the complex projective space of dimension $2d + 1$). A *vector field above* f is a map $w : \widehat{\mathbb{C}} \to T\widehat{\mathbb{C}}$ such that $\pi \circ w = f$, where $\pi : T\widehat{\mathbb{C}} \to \widehat{\mathbb{C}}$ is the canonical projection onto the base of the tangent bundle of the sphere. We say that a continuous vector field $v : \widehat{\mathbb{C}} \to T\widehat{\mathbb{C}}$ is a *deformation* of f if $\delta v = f'v - v \circ f$ is a *holomorphic vector field above* f. Note that since

$$\overline{\partial}(\delta v) \;=\; f'\,\overline{\partial}v - \overline{\partial}v \circ f\,\overline{f'}$$

such a v is a deformation of f if and only if the Beltrami differential $\mu = \overline{\partial}v$ is f-invariant. We say that v is a *trivial* deformation of f if $f^*(v) = v$ (its pull-back as a vector field) or equivalently if $\delta v = 0$.

Lemma 4.6.3 *If $v : \widehat{\mathbb{C}} \to T\widehat{\mathbb{C}}$ is continuous and yields a trivial deformation of f then v is identically zero on the Julia set of f.*

Proof If $\delta v = 0$ then

$$f'(z)v(z) \;=\; v(f(z))$$

for all $z \in \widehat{\mathbb{C}}$. Hence, by the chain rule,

$$\left(f^k\right)'(z)\,v(z) \;=\; v\left(f^k(z)\right)$$

for all $z \in \widehat{\mathbb{C}}$ and all $k \geq 0$. In particular, if $z = p$ is a k-periodic point of f, we have

$$\left(f^k\right)'(p)\,v(p) \;=\; v(p)\,.$$

If $(f^k)'(p) \neq 1$, which is certainly the case if p is repelling, then $v(p) = 0$. Thus v vanishes at the repelling periodic points, and since these are dense in the Julia set of f, the result follows. $\qquad\square$

Following McMullen [McM4], we denote by $M_f(\widehat{\mathbb{C}})$ the vector space of f-invariant Beltrami differentials on the Riemann sphere. The Ahlfors–Bers theorem yields an almost natural map from $M_f(\widehat{\mathbb{C}})$ to the tangent space at f of the space of degree-d rational maps. Indeed, if $\mu \in M_f(\widehat{\mathbb{C}})$ then $\|t\mu\|_\infty < 1$ for all t in a small disk about the origin. Let $\phi_t : \widehat{\mathbb{C}} \to \widehat{\mathbb{C}}$ be the unique normalized solution to $\overline{\partial}\phi_t = t\mu\partial\phi_t$. By the Ahlfors–Bers theorem, both ϕ_t and ϕ_t^{-1} vary holomorphically with t, so that $f_t = \phi_t \circ f \circ \phi_t^{-1} \in \mathrm{Rat}_d(\widehat{\mathbb{C}})$ varies holomorphically with t also. We can now take

$$w \;=\; \frac{d}{dt}\bigg|_{t=0} f_t \in T_f\mathrm{Rat}_d(\widehat{\mathbb{C}})\,.$$

Thus we have a well-defined map

$$L : M_f(\widehat{\mathbb{C}}) \to T_f \mathrm{Rat}_d(\widehat{\mathbb{C}}) .$$

This map can be alternatively described as follows. Given $\mu \in M_f(\widehat{\mathbb{C}})$, find a continuous vector field v such that $\overline{\partial} v = \mu$, using, say theorem 4.5, normalized in such a way that $v(0) = v(1) = v(\infty) = 0$. Then take $w = L(\mu) = \delta v$. This w is holomorphic because μ is f-invariant. It is clear from this alternative definition that L is linear.

Now the crucial idea is that $M_f(\widehat{\mathbb{C}})$ is a huge space, typically infinite dimensional, whereas $T_f \mathrm{Rat}_d(\widehat{\mathbb{C}})$ has complex dimension $2d + 1$, so that L cannot be injective on any subspace of dimension $2d + 2$ or higher. Hence, Sullivan's theorem will follow by contradiction if we can build a sufficiently large space of quasiconformal deformations of f restricted to which L is injective. The key to constructing such large space is the following lemma. Let us agree to call a Beltrami differential $\mu \in M(\mathbb{D})$ *trivial* if there exists a continuous vector field v such that $\overline{\partial} v = \mu$ and $v = 0$ on $\partial\mathbb{D}$.

Lemma 4.6.4 *There exists an infinite-dimensional space $V \subset M(\mathbb{D})$ of compactly supported Beltrami differentials with the property that $\mu \in V$ is trivial if and only if $\mu = 0$.*

Proof For each n-tuple $(a_1, a_2, \ldots, a_n) \in \mathbb{C}^n$, let

$$\mu_{a_1,a_2,\ldots,a_n}(z) = \begin{cases} \displaystyle\sum_{k=1}^{n} a_k \overline{z}^{k-1} & \text{if } |z| \leq \frac{1}{2}, \\ 0 & \text{if } |z| > \frac{1}{2}. \end{cases}$$

The set of all $\mu_{a_1,a_2,\ldots,a_n}$ defined in this way, as we vary n and the n-tuple $(a_1, a_2, \ldots, a_n)$, is obviously an infinite-dimensional vector space, which we denote by V. We see at once that

$$\mu_{a_1,a_2,\ldots,a_n} = \overline{\partial} v_{a_1,a_2,\ldots,a_n} ,$$

where $v_{a_1,a_2,\ldots,a_n}$ is the continuous vector field given by

$$v_{a_1,a_2,\ldots,a_n}(z) = \begin{cases} \displaystyle\sum_{k=1}^{n} \frac{a_k}{k} \overline{z}^{k} & \text{if } |z| \leq \frac{1}{2}, \\ \displaystyle\sum_{k=1}^{n} \frac{a_k}{k4^k} z^{-k} & \text{if } |z| > \frac{1}{2}. \end{cases}$$

Now suppose that $\mu = \mu_{a_1,a_2,\ldots,a_n} \in V$ is trivial. Then $\overline{\partial}w = \mu$ for some w with $w = 0$ on $\partial\mathbb{D}$. Hence $\overline{\partial}(v_{a_1,a_2,\ldots,a_n} - w) = 0$, so $v_{a_1,a_2,\ldots,a_n} - w$ is holomorphic on the disk. Since w vanishes at the boundary, it follows that the restriction of v to $\partial\mathbb{D}$ has a holomorphic extension to $\mathbb{D}$. Since $v|\partial\mathbb{D}$ is a polynomial in z^{-1}, we deduce that $a_1 = a_2 = \cdots = a_n = 0$, and therefore $\mu = 0$. $\qquad\square$

This lemma yields the following result.

Lemma 4.6.5 *Let $U \subseteq \widehat{\mathbb{C}}$ be a simply connected domain, and let $\varphi : \mathbb{D} \to U$ be a Riemann map. Suppose that v is a vector field on the Riemann sphere that vanishes on ∂U and is such that $\mu = \overline{\partial}v|U$ is compactly supported. Then the pull-back $\varphi^*(v)$ of v to the unit disk vanishes at $\partial\mathbb{D}$, and the pull-back $\varphi^*(\mu)$ is trivial in $M(\mathbb{D})$.* $\square$

We leave the proof of this lemma as an exercise for the reader. We now have all the necessary ingredients to finish the proof of Sullivan's no-wandering-domains theorem.

Proof of theorem 4.6.1. By lemma 4.6.2, it suffices to rule out wandering (topological) disks. By contradiction, suppose that U is a wandering disk for f and let $\varphi : \mathbb{D} \to U$ be a Riemann map. Then φ^* establishes an isomorphism between $M(\mathbb{D})$ and $M(U)$, and in particular the space V of lemma 4.6.4 injects into $M(U)$. Now, $M(U)$ injects into $M_f(\widehat{\mathbb{C}})$ by dynamics. Indeed, from a given $\nu \in M(U)$ we get an element $\tilde{\nu} \in M_f(\widehat{\mathbb{C}})$ by first spreading ν to the whole grand orbit of U, namely

$$[U] = \bigcup_{m \geq 0}\bigcup_{n \geq 0} f^{-m}\left(f^n(U)\right) .$$

This is done by pushing ν to each forward image $f^n(U)$ via the *injective* map $f^n : U \to f^n(U)$ and then taking the pull-backs via the various inverse branches f^{-m}. We complete the definition of $\tilde{\nu}$ by setting it equal to zero on $\widehat{\mathbb{C}} \setminus [U]$. Thus we have an isomorphism $\iota : V \to \tilde{V} \subset M_f(\widehat{\mathbb{C}})$. We now claim that $L|\tilde{V}$ is injective. To see why, suppose that $\tilde{\mu} \in \tilde{V}$ is such that $L(\tilde{\mu}) = 0$; in other words, suppose that $\tilde{\mu}$ yields a trivial deformation of f. Then there exists a vector field v on the sphere with $\overline{\partial}v = \tilde{\mu}$ and $\delta v = 0$. By lemma 4.6.3, v vanishes identically on the Julia set of f and in particular $v|\partial U$ is equal to zero. By lemma 4.6.5, $\mu = \iota^{-1}(\tilde{\mu}) \in V$ is a trivial Beltrami differential on the unit disk and therefore $\mu = 0$ by lemma 4.6.4. Hence $\tilde{\mu} = 0$ also, and $L|\tilde{V}$ is injective as claimed. This is impossible, however, because the domain $\tilde{V}$ is infinite dimensional while the range $T_f\mathrm{Rat}_d(\widehat{\mathbb{C}})$ is finite dimensional. This contradiction shows that f has no wandering disks, and this completes the proof.   □

4.7 Boundary behavior of quasiconformal maps

We saw in proposition 4.3.16 that a quasiconformal homeomorphism of the unit disk extends to a quasiconformal homeomorphism of the Riemann sphere. In particular it defines a homeomorphism of the unit circle. Similarly, a quasiconformal homeomorphism of the upper half-plane $\mathbb{H}$ extends to a quasiconformal homeomorphism of the Riemann sphere and therefore defines a homeomorphism of the real line. In this section, we will describe a characterization of all those homeomorphisms of the real line or of the unit circle that occur as boundary values of quasiconformal homeomorphisms.

Definition 4.7.1 *A homeomorphism* $\phi\colon \mathbb{R} \to \mathbb{R}$ *is M-quasisymmetric if*

$$\frac{1}{M} \leq \frac{|\phi(x+t) - \phi(x)|}{|\phi(x) - \phi(x-t)|} \leq M \,,$$

for all $x \in \mathbb{R}$ and all $t \neq 0$; the smallest M with this property is called the quasisymmetric distortion of ϕ.

We define quasisymmetric homeomorphisms of the circle in a similar fashion: the image of the midpoint of any arc on the circle cannot be too close to the boundary of the image arc, i.e. it splits the image into two sub-arcs and the ratio of the lengths of these two sub-arcs is bounded away from zero and infinity.

Proposition 4.7.2 *Given $K > 1$, there exists an $M > 1$ that depends only on K such that the map determined of the boundary values of any K-quasiconformal homeomorphism of the upper half-plane is an M-quasisymmetric homeomorphism of the real line.*

Proof Let $x - t, x, x + t$ be any three symmetric points on the real line and consider the quadrilaterals $\mathbb{H}(x - t, x, x + t, \infty)$. They are all conformally equivalent to $\mathbb{H}(-1, 0, 1, \infty)$ by a Möbius transformation and, therefore, they have the same modulus. The image of this quadrilateral is a quadrilateral whose modulus is bounded and bounded away from zero by constants that depend only on K. However, the modulus of the quadrilateral $\mathbb{H}(y, y - A, y + B, \infty)$ is bounded away from zero and infinity if and only if A/B is bounded away from zero and infinity. The quasisymmetry condition follows.   □

The converse, stated below, was proved by Beurling and Ahlfors in [BA].

Theorem 4.7.3 (Beurling–Ahlfors) *Given $L > 1$, there exists a $K > 1$ that depends only on L such that any S-quasisymmetric homeomorphism $\phi \colon \mathbb{R} \to \mathbb{R}$ has a continuous extension to a K-quasiconformal homeomorphism $\phi \colon \mathbb{H} \to \mathbb{H}$.*

The proof of the above theorem uses an explicit formula for the extension. If we write $\Phi(x, y) = u(x, y) + iv(x, y)$ then

$$u(x,y) = \frac{1}{2y} \int_{-y}^{y} \phi(x+t)\, dt,$$

$$v(x,y) = \frac{1}{2y} \int_{0}^{y} (\phi(x+t) - \phi(x-t))\, dt.$$

From this formula one can prove that Φ extends ϕ continuously. By a simple change of coordinates we get

$$u(x,y) = \frac{1}{2y} \int_{x-y}^{x+y} \phi(t)\, dt,$$

$$v(x,y) = \frac{1}{2y} \left(\int_{x}^{x+y} \phi(t)\, dt - \int_{x-y}^{x} \phi(t)\, dt \right),$$

which shows that Φ is differentiable. We refer to [A1, p. 69] for a complete proof of the Beurling–Ahlfors theorem.

4.7.1 The Douady–Earle extension

Next we discuss another important formulation of the extension theorem of quasisymmetric homeomorphisms by Douady and Earle, in [DE]. They proved the existence of an extension map $\mathcal{E}$, from the space of orientation-preserving homeomorphisms of the circle into the space of homeomorphisms of the closure of the unit disk $\mathbb{D}$, that is conformally natural in the sense that $\mathcal{E}$ is *equivariant* with respect to the action of the Möbius group.

Let us consider the action of the group $M(\mathbb{D})$, of Möbius transformations preserving the unit disk, on the closure of the disk: $M(\mathbb{D}) \times \overline{\mathbb{D}} \to \overline{\mathbb{D}}$, $(g, z) \mapsto g(z)$. This restricts to an action on the circle S^1, which induces an action on various spaces as follows:

(1) on the space $\mathcal{P}(S^1)$ of probability measures on S^1 by pushing forward: $(g, \mu) \mapsto g^* \mu$, where $g^* \mu(A) = \mu(g^{-1}(A))$;
(2) on the spaces of continuous functions $C^0(S^1)$ and $C^0(\overline{\mathbb{D}})$ by gf $(x) = f \circ g^{-1}$;
(3) on the space of continuous vector fields on $\mathbb{D}$, $\chi^0(\mathbb{D})$, by $gV(t) = Dg\,(g^{-1}(t))V(g^{-1}(t))$ and on in the spaces of homeomorphisms $\mathrm{Homeo}(S^1)$ and $\mathrm{Homeo}(\overline{\mathbb{D}})$ by composition, $(g, \phi) \mapsto \phi \circ g^{-1}$, and also by $(g, \phi) \mapsto g \circ \phi$.

If a group G acts in two spaces X, Y, a mapping $F \colon X \to Y$ is equivariant if $F(gx) = gF(x)$. As an example, we may consider the action of the Möbius group in the space of functions; then there is a unique equivariant function $H \colon C^0(S^1) \to C^0(\overline{\mathbb{D}})$ that is linear, preserves the positive cone of non-negative functions and maps constant functions to constant functions. It maps continuous maps on the circle into harmonic maps on the disk.

4.8 Polynomial-like maps

Let us now move to another very important application of the measurable Riemann mapping theorem, to the theory of *polynomial-like mappings* created by A. Douady and J. Hubbard [DH1, DH2]; see figures 4.1 and 4.2.

Definition 4.8.1 *A polynomial-like map of degree d is a proper holomorphic branched covering map $f \colon U \to V$ of degree d, where U and V are Jordan domains and U is compactly contained in V. The filled-in Julia set of f is the compact set $K(f) = \{z \in U : f^n(z) \in U, \forall n \in \mathbb{N}\}$; the boundary $J(f) = \partial K(f)$ is called the Julia set of f.*

Example 4.8.2 Let f be a polynomial of degree $d \geq 2$. If V is a disk centered at the origin and of sufficiently large radius R, then f^{-1} is a topological disk compactly contained in V. The restriction of f to U is therefore a polynomial-like map of degree d.

Definition 4.8.3 *Two polynomial-like maps $f_i \colon U_i \to V_i$, $i = 1, 2$, are hybrid equivalent if there exists a quasiconformal mapping $h \colon V_1 \to V_2$ such that the restriction of h to U_1 conjugates f_1 with f_2 and such that $\bar{\partial} h = 0$ at Lebesgue-almost-all points in the filled-in Julia set of f_1.*

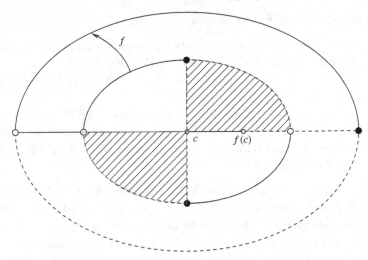

Fig. 4.1. A quadratic-like map with critical point c.

The following theorem, presented by Douady and Hubbard in [DH2], shows the strength of the measurable Riemann mapping theorem.

Theorem 4.8.4 (Straightening theorem) *Let $f\colon U \to V$ be a polynomial-like map of degree $d \geq 2$. Then f is hybrid equivalent to the restriction of a polynomial g of degree d to some Jordan domain. If the filled-in Julia set of f is connected then the polynomial g is uniquely determined up to conjugacy by an affine map.*

Proof By shrinking V a little if necessary, we may take the boundaries of V and U to be smooth curves (in fact real analytic). Let U' be the disk of radius 2 centered at the origin and V' be the disk of radius 2^d that is the image of U' by the map $g : z \mapsto z^d$. Let h_0 be a diffeomorphism of the closed ring domain $\overline{(V \setminus U)}$ onto $\overline{(V' \setminus U')}$ that conjugates f and g in the boundary, i.e. $g \circ h_0(z) = h_0 \circ f(z)$ if $z \in \partial U$. Let S be the quotient space of the disjoint union of V with $\widehat{\mathbb{C}} \setminus U'$ by the equivalence relation that identifies $z \in V \setminus U$ with $h_0(z) \in V' \setminus U'$. Clearly S is diffeomorphic to the sphere, and $F\colon S \to S$ defined by $F(z) = f(z)$ if $z \in U$ and $F(z) = z^d$ if $z \in \widehat{\mathbb{C}} \setminus U'$ is a smooth map. On S, let us consider the field of ellipses $z \mapsto E(z)$ defined as follows: for $z \in \widehat{\mathbb{C}} \setminus U'$, $E(z)$ is a round circle; for $z \in V \setminus U$, $E(z)$ is the pull-back by the derivative $dh_0(z)$ of the circle $E(h_0(z))$; also, $E(z)$ is the pull-back of $E(f^n(z))$ by the derivative

$Df^n(z)$ if $f^n(z) \in V \setminus U$; and finally, $E(z)$ is a circle if $f^n(z) \in U$ for
all $n \in \mathbb{N}$. This is clearly a measurable field of ellipses; the eccentricity
is 1 in the complement of V and it is uniformly bounded in the fundamental domain $V \setminus U$ because h_0 is a C^1 diffeomorphism of the closure
of the fundamental domain, which is compact. The eccentricities of the
pull-back of these ellipses by iterates of f remain bounded by the same
constant, because f^n is holomorphic. By the measurable Riemann mapping theorem, there exists a quasiconformal homeomorphism $h: S \to \widehat{\mathbb{C}}$
such that the field of ellipses just constructed is the pull-back of the
field of circles. Hence the mapping $G = h \circ F \circ h^{-1}$ preserves the field of
circles and, therefore, is a holomorphic map of degree d. Furthermore, G
has a fixed point, $h(\infty)$, that is totally invariant. This implies that the
Möbius transformation that maps this point to ∞ conjugates G with a
polynomial of degree d. The restriction to V of the composition of the
Möbius transformation with h is clearly a hybrid equivalence.

It remains to prove that if two polynomials are hybrid equivalent
and their filled-in Julia sets are connected then they are conformally
equivalent. Let f, g be two polynomial-like maps with connected filled-in

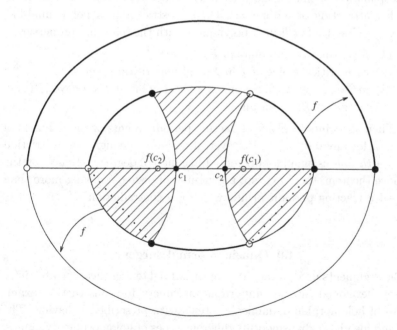

Fig. 4.2. A polynomial-like map of degree with critical points c_1 and c_2.

Julia sets, and let ϕ be a hybrid conjugacy between f and g. Since the filled-in Julia set $K(f)$ is connected, $\widehat{\mathbb{C}} \setminus K(f)$ is simply connected. By the Riemann mapping theorem, there exists a unique univalent map $\psi \colon \widehat{\mathbb{C}} \setminus K(f) \to \widehat{\mathbb{C}} \setminus \overline{\mathbb{D}}$ that fixes ∞ and whose derivative at ∞ is real and positive. This mapping is a holomorphic conjugacy between f and $z \mapsto z^d$. Since the same holds for g, we deduce that there exists a univalent map $\theta \colon \widehat{\mathbb{C}} \setminus K(f) \to \widehat{\mathbb{C}} \setminus K(g)$ that conjugates f and g. Gluing this map to the hybrid conjugacy, we get a quasiconformal map $h_0 \colon \widehat{\mathbb{C}} \to \widehat{\mathbb{C}}$ such that the restriction of h_0 to a neighborhood of ∞ is a holomorphic conjugacy and the restriction of h_0 to $K(f)$ has zero dilatation almost everywhere. Thus, h_0 is almost everywhere conformal and therefore conformal. $\qquad\square$

Corollary 4.8.5 *If f is a polynomial of degree d then the number of periodic orbits of f that are either attracting or indifferent is less than or equal to $d - 1$.*

Proof If this were not the case, we could select a finite subset E of the plane consisting of more than $d - 1$ periodic orbits of f that are either attracting or indifferent. Let us restrict f to a polynomial-like map $f \colon U \to V$. Let h be a polynomial with the following properties:

(1) h vanishes at all points of E;
(2) each critical point of f in E is also a critical point of h;
(3) for each $z \in E$ that is not a critical point of f we have $|f'(z) + th'(z)| < |f'(z)|$ for all $0 < t \le 1$.

Then all points in E are attracting periodic points of the polynomial $f + th$ for any $0 < t \le 1$. However, if t is small enough, the restriction of $f + th$ to U is again a polynomial-like map of degree d. Hence, by the above theorem, there exists a polynomial of degree d having more than $d - 1$ attracting periodic orbits, which is a contradiction. $\qquad\square$

4.9 Quasiconformal surgery

The arguments used in the proof of the straightening theorem 4.8.4 have been developed into an important machinery for constructing examples of holomorphic dynamical systems with prescribed behavior. The rough idea is to glue smoothly different types of holomorphic dynamics, producing a non-holomorphic dynamical system with an invariant

Beltrami differential that can be straightened using the measurable Riemann mapping theorem. This construction is now called *quasiconformal surgery*, and was used by Shishikura to prove the following result.

Theorem 4.9.1 (Shishikura) *If f is a rational map of degree d then*

$$n_A + n_P + n_I + 2n_H \leq 2d - 2,$$

where n_A is the number of attracting cycles, n_P is the number of cycles of parabolic domains, n_I is the number of cycles of irrationally indifferent periodic points and n_H is the number of cycles of Herman rings.

For a proof, the reader should consult the original reference, [Sh1].

Although we will not prove the above theorem, we will establish a related surgery result, also due to Shishikura, presented in the same article, [Sh1]. The theorem will give another proof of the existence of rational maps having Herman rings. But before we state it, we need a version of what is known as the *fundamental lemma of quasiconformal surgery*. Let us say that a branched covering of the Riemann sphere is quasiregular if it is the composition of a rational map with a quasiconformal homeomorphism. Let us also agree to call a homeomorphism *anti-quasiconformal* if it is the composition of a quasiconformal homeomorphism with an anti-conformal map (such as complex conjugation).

Lemma 4.9.2 *Let $G : \widehat{\mathbb{C}} \to \widehat{\mathbb{C}}$ be a quasiregular map, and let $E \subseteq \widehat{\mathbb{C}}$ be an open set. Suppose that*

(i) *we have $G(E) \subseteq E$;*
(ii) *there exists a homeomorphism $\Psi : \widehat{\mathbb{C}} \to \widehat{\mathbb{C}}$ that is (anti-)quasiconformal and such that $\Psi \circ G \circ \Psi^{-1}|_{\Psi(E)}$ is holomorphic;*
(iii) *there exists $N \geq 0$ such that G is holomorphic in $\widehat{\mathbb{C}} \setminus G^{-N}(E)$.*

Then there exists a quasiconformal homeomorphism $H : \widehat{\mathbb{C}} \to \widehat{\mathbb{C}}$ such that $H \circ G \circ H^{-1}$ is a rational map.

Proof Assume that Ψ is quasiconformal, and let μ_Ψ be the Beltrami coefficient of Ψ. Define a new Beltrami coefficient μ as follows. First set $\mu = \mu_\Psi$ in E. Then, given $z \in \widehat{\mathbb{C}} \setminus E$, let $\mu(z) = (G^n)^*\mu(z)$ (the pull-back of μ under G^n), where n is the smallest positive integer such that $G^n(z) \in E$, if there is such an n, and let $\mu(z) = 0$ otherwise. Then $\mu \in L^\infty(\widehat{\mathbb{C}})$. By condition (iii), in the composition making up the iterate G^n, all but a bounded number $\leq N$ of factors are conformal; this implies that $\|\mu\|_\infty < 1$. Moreover, conditions (i)–(iii) above entail that μ is G-invariant. Hence, solving the Beltrami equation for μ, we get a quasiconformal homeomorphism $H : \widehat{\mathbb{C}} \to \widehat{\mathbb{C}}$ such that $H \circ G \circ H^{-1}$ is holomorphic and therefore a rational map. The same proof works, *mutatis mutandis*, if Ψ is anti-quasiconformal. $\qquad\square$

We remark that, in the above lemma, the map $H \circ \Psi^{-1}$ is either conformal or anti-conformal in $\Psi(E)$, depending on whether Ψ is quasiconformal or anti-quasiconformal respectively.

Theorem 4.9.3 (Shishikura) *Let $0 < \theta < 1$ be an irrational number, and suppose that there is a rational map f of degree d having a Siegel disk restricted to which f is conjugate to the rotation $R_\theta : z \mapsto e^{2\pi i\theta}z$. Then there exists a rational map of degree $2d$ having a Herman ring with the same rotation number.*

Proof Let V_f be the Siegel disk of f. We may assume that $0 \in V_f$ and that $f(0) = 0$. Let $g \in \mathrm{Rat}_d(\widehat{\mathbb{C}})$ be the rational map $g = c \circ f \circ c^{-1}$, where $c(z) = \overline{z}$ (of course, $c = c^{-1}$). Then g also has a Siegel disk $V_g = c(V_f) \ni 0$, with $g(0) = 0$, and $g|_{V_g}$ is conjugate to $R_{-\theta} : z \mapsto e^{-2\pi i\theta}z$. Let $h_f : V_f \to \mathbb{D}$ be a conformal conjugacy between $f|_{V_f}$ and $R_\theta|_\mathbb{D}$. Likewise, let $h_g : V_g \to \mathbb{D}$ be a conformal conjugacy between $g|_{V_g}$ and $R_{-\theta}|_\mathbb{D}$; in fact, we can take $h_g = h_f \circ c^{-1}$.

Fix $0 < r < 1$, and let $C_r = \{z : |z| = r\} \subset \mathbb{D}$. Take an open annulus A around C_r, with $\overline{A} \subset \mathbb{D}$, that is symmetric under inversion about C_r. In other words, $\psi(A) = A$, where $\psi(z) = r^2/z$. Note that $\psi(C_r) = C_r$ and that ψ satisfies

$$\psi \circ R_\theta(z) = \frac{r^2}{e^{2\pi i\theta}z} = R_{-\theta} \circ \psi(z) \ .$$

Next, consider the topological annuli $A_f = h_f^{-1}(A) \subset V_f$ and $A_g = h_g^{-1}(A) \subset V_g$. We have a conformal map

$$\phi = h_g^{-1} \circ \psi \circ h_f : A_f \to A_g \ .$$

This maps conjugates $f|_{A_f}$ to $g|_{A_g}$. The invariant curves $\gamma_f = h_f^{-1}(C_r)$ and $\gamma_g = h_g^{-1}(C_r)$ correspond to each other via this conjugacy, i.e. $\phi(\gamma_f) = \gamma_g$. Moreover, if $D_f \subset V_f$ and $D_g \subset V_g$ are the topological disks bounded by γ_f and γ_g respectively, we see that ϕ maps $A_f \cap D_f$ onto $A_g \cap (\widehat{\mathbb{C}} \setminus D_g)$ and $A_f \cap (\widehat{\mathbb{C}} \setminus D_f)$ onto $A_g \cap D_g$.

Now let $B_f \subset \overline{B_f} \subset A_f$ be another topological annulus that is invariant under f and contains γ_f, and let $B_g = \phi(B_f)$. We define a quasiconformal homeomorphism $\Phi : \widehat{\mathbb{C}} \to \widehat{\mathbb{C}}$ extending $\phi|_{B_f}$ by letting Φ be a conformal map from each component of $\widehat{\mathbb{C}} \setminus A_f$ onto the corresponding component of $\widehat{\mathbb{C}} \setminus A_g$, using for instance the Riemann mapping theorem, and interpolating by diffeomorphisms in any way we like in each annular component of $A_f \setminus B_f$.

Next, let $G : \widehat{\mathbb{C}} \to \widehat{\mathbb{C}}$ be the map defined by

$$
G(z) = \begin{cases} f(z) & \text{if } z \in \widehat{\mathbb{C}} \setminus D_f \,, \\ \\ \Phi^{-1} \circ g \circ \Phi(z) & \text{if } z \in D_f \,. \end{cases}
$$

Since

$$
f(z) = \phi^{-1} \circ f \circ \phi(z) = \Phi^{-1} \circ g \circ \Phi(z)
$$

for all $z \in A_f \supset \partial D_f (= \gamma_f)$, it follows that G is a branched covering of the Riemann sphere and in fact a quasiregular map. This map G has twice as many branch points as f: for each branch point it has in $\widehat{\mathbb{C}} \setminus V_f$, it has a corresponding branch point in $D_f \cap (\widehat{\mathbb{C}} \setminus A_f)$ (coming from g) with the same multiplicity. Therefore the topological degree of G is equal to $2d$.

The next step is to deform quasiconformally G into a rational map. This is where we use lemma 4.9.2 above. Its hypotheses are satisfied if we take $E = A_f$, $\Psi = \Phi^{-1}$ and $N = 1$, as the reader can easily check. Let $H : \widehat{\mathbb{C}} \to \widehat{\mathbb{C}}$ be the quasiconformal homeomorphism whose existence is guaranteed by that lemma. Then $F = H \circ G \circ H^{-1}$ is a rational map, with the same degree ($2d$) as G.

Finally, let $W = H(A_f)$. This is a topological annulus with $F(W) = W$, so W is contained in some component U of the Fatou set of F. Since $F|_W$ is clearly conjugate to the rotation R_θ, we see that U is either a Siegel disk or a Herman ring for F. In particular, $F|_U$ is injective. If U is a Siegel disk then it contains one of the two components of $\widehat{\mathbb{C}} \setminus W$; but this is impossible because both such components contain critical points of F. Therefore U is a Herman ring, with rotation number θ, and the theorem is proved. $\qquad\square$

Exercises

4.1   Let $f : U \to V$ and $g : V \to W$ be quasiconformal diffeomorphisms between domains U, V, W in the plane, and let μ_f and

μ_g be their Beltrami differentials. Show that $g \circ f : U \to W$ is quasiconformal, with Beltrami coefficient $\mu_{g \circ f}$ given by

$$\mu_{g \circ f}(z) = \frac{\mu_f(z) + \mu_g(f(z))\theta_f(z)}{1 + \overline{\mu_f(z)}\mu_g(f(z))\theta_f(z)} \, ,$$

where $\theta_f(z) = \overline{f_z(z)}/f_z(z)$.

4.2   Using the result of exercise 4.1, prove the following assertions made at the beginning of this chapter.

    (a) The inverse of a K-quasiconformal diffeomorphism is K-quasiconformal.

    (b) The composition of a K-quasiconformal diffeomorphism with a K'-quasiconformal diffeomorphism is KK'-quasiconformal.

4.3   Let Γ_1 and Γ_2 be two families of piecewise C^1 curves in the plane, and suppose that the traces of the curves of Γ_1 and those of Γ_2 lie in disjoint measurable sets. Prove that

$$\frac{1}{\lambda(\Gamma_1)} + \frac{1}{\lambda(\Gamma_2)} \leq \frac{1}{\lambda(\Gamma_1 \cup \Gamma_2)} \, .$$

4.4   Let $\alpha > 0$, and consider the map $f_\alpha : \mathbb{C} \to \mathbb{C}$ given by $f_\alpha(re^{i\theta}) = r^\alpha e^{i\theta}$. Show that f_α is a quasiconformal homeomorphism, and compute its dilatation.

4.5   Is there a quasiconformal homeomorphism mapping the complex plane into the upper half-plane? Explain.

4.6   Show that if $\varphi \in C_0^\infty(\mathbb{C})$ then $\|S(\varphi)\|_2 = \|\varphi\|_2$, where S is the Beurling transform.

4.7   Let $\phi \in L^p(\mathbb{C})$ with $p > 2$ and suppose that ϕ has compact support. Define

$$\psi(z) = \frac{1}{2\pi i} \iint_{\mathbb{C}} \frac{\phi(\zeta)}{\zeta - z} \, d\zeta \wedge d\overline{\zeta} \, .$$

Prove that ψ is a solution to $\overline{\partial}\psi = \phi$ in the distributional sense, working through the following steps.

    (a) First suppose that $\phi \in C_0^\infty(\mathbb{C})$, and prove the required result in this case with the help of Pompeiu's formula (note that $\overline{\partial}\phi$ also has compact support and therefore $\overline{\partial}\phi \in L^p(\mathbb{C})$ for all p).

(b) If now ϕ still has compact support but is merely in $L^p(\mathbb{C})$, let $\phi_n \in C_0^\infty(\mathbb{C})$, $n \geq 1$, be a smoothing sequence converging to ϕ in the L^p sense and define

$$\psi_n(z) = \frac{1}{2\pi i} \iint_{\mathbb{C}} \frac{\phi_n(\zeta)}{\zeta - z} \, d\zeta \wedge d\bar{\zeta} \ .$$

Show that $\psi_n \to \psi$ uniformly on compact sets (note that one can take a smoothing sequence $\{\phi_n\}$ such that $\operatorname{supp} \phi_n \subset D(0, R)$ for all n, for some sufficiently large $R > 0$).

(c) Check that $\bar{\partial}\psi_n = \phi_n$ for all n, so that for every test function $\varphi \in C_0^\infty(\mathbb{C})$ one has

$$\iint_{\mathbb{C}} \psi_n \, \bar{\partial}\varphi \, d\zeta \wedge d\bar{\zeta} = - \iint_{\mathbb{C}} \phi_n \varphi \, d\zeta \wedge d\bar{\zeta} \ .$$

(d) Deduce from (c) that for every test function $\varphi \in C_0^\infty(\mathbb{C})$ one has

$$\iint_{\mathbb{C}} \psi \, \bar{\partial}\varphi \, d\zeta \wedge d\bar{\zeta} = - \iint_{\mathbb{C}} \phi \varphi \, d\zeta \wedge d\bar{\zeta} \ ,$$

and therefore $\bar{\partial}\psi = \phi$ in the sense of distributions.

4.8   Show that ψ in the previous exercise satisfies a Hölder condition, with exponent $1 - 2/p$ and Hölder constant $C = C(p, \|\phi\|_p, R)$, where $R > 0$ is such that $\operatorname{supp} \phi \subset D(0, R)$. Prove also that ψ is holomorphic in a neighborhood of ∞ and that $\psi(z) = O(1/z)$ as $|z| \to \infty$.

4.9   Let $\Gamma \subset PSL(2, \mathbb{C})$ be a discrete subgroup of the Möbius group, and let $\mu \in L^\infty(\widehat{\mathbb{C}})$ with $\|\mu\|_\infty < 1$ be Γ-invariant, in the sense that

$$\mu(\gamma(z)) \frac{\overline{\gamma'(z)}}{\gamma'(z)} = \mu(z)$$

for all $z \in \widehat{\mathbb{C}}$ and all $\gamma \in \Gamma$. If $f : \widehat{\mathbb{C}} \to \widehat{\mathbb{C}}$ is a solution of the Beltrami equation $\bar{\partial}f = \mu \, \partial f$, show that $f \circ \gamma = \gamma \circ f$ for all $\gamma \in \Gamma$.

4.10   Let $f : \mathbb{R} \to \mathbb{R}$ be a piecewise differentiable homeomorphism such that $K^{-1} \leq f'(x) \leq K$ for all x, where $K > 1$ is a constant. Show that the extension $F : \mathbb{H} \to \mathbb{H}$ of f to the upper half-plane given by $F(x, y) = f(x) + iy$ is a K-quasiconformal homeomorphism.

4.11   (See [dF3, p. 1015]) For each $0 \leq \theta < 1$, let $E_\theta : \mathbb{C} \to \mathbb{C}$ be the
entire map given by

$$E_\theta(z) \;=\; z + \theta - \frac{1}{2\pi} \sin(2\pi z) \;.$$

(a) Check that $E_\theta \circ T = T \circ E_\theta$, and deduce that E_θ is the
lift to the plane of a holomorphic map $f_\theta : \mathbb{C}^* \to \mathbb{C}^*$ via
the exponential covering map.

(b) Verify that each $f_\theta | S^1$ maps the unit circle onto itself and
that it is a critical circle homeomorphism with a unique
critical point at $z = 1$.

(c) Prove that the family $\{f_\theta\}$ is *topologically complete*, in
other words that if a holomorphic map $f : \mathbb{C}^* \to \mathbb{C}^*$,
normalized to have its critical point at $z = 1$ and symmetric with respect to inversion about the unit circle, is
topologically conjugate to a member of the family then
f is a member also.

(d) Let $\rho(\theta)$ be the rotation number of $f_\theta | S^1$. Show that if
$\rho(\theta)$ is irrational then f_θ admits no non-trivial invariant
Beltrami differential entirely supported in its Julia set
$J(f_\theta)$, by working through the following steps.

  (1) First, using Montel's theorem, show that

  $$J(f_\theta) \subseteq \overline{\bigcup_{n \geq 0} f_\theta^{-n}(1)} \;.$$

  (2) Let μ be such an f-invariant Beltrami differential
  with support in $J(f_\theta)$ and, for each sufficiently
  small real t, let $h_t : \widehat{\mathbb{C}} \to \widehat{\mathbb{C}}$ be the unique solution
  to $\overline{\partial} h_t = (t\mu)\partial h_t$, normalized so as to fix $0, 1, \infty$.
  Let $f_t = h_t \circ f \circ h_t^{-1}$. Using part (c) of the exercise,
  deduce that $f_t = f_{\theta(t)}$ for some $\theta(t)$.

  (3) Using (2) and certain well-known facts about the
  function $\theta \mapsto \rho(\theta)$, show for all t that $f_{\theta(t)} = f_\theta$,
  that h_t commutes with f_θ and therefore that h_t
  permutes the elements of the discrete set $f^{-n}(1)$,
  for each $n \geq 0$.

  (4) Using (1) and (3) and applying the Ahlfors–Bers
  theorem to $t \to h_t(z)$, deduce that h_t is the identity for every t, so that μ must vanish almost
  everywhere.

5

Holomorphic motions

When studying the structural stability of complex dynamical systems, we are naturally led to consider families of holomorphic systems depending holomorphically on a given set of parameters. While the concept of a holomorphically varying family of Riemann surfaces had already been around for some time in the theory of moduli spaces, it was not until the work of Mañé, Sad and Sullivan [MSS] that the idea of *holomorphic motion* made its appearance in complex dynamics. In this chapter, we shall present some of the main facts about holomorphic motions, including the original motivation behind [MSS], namely the structural stability properties of rational maps. This study will be carried out with the help of the Bers–Royden theorem, a deep result that will be presented in detail.

5.1 Holomorphic functions in Banach spaces

We begin with some preliminary facts concerning holomorphic functions in Banach spaces. Let X be a *complex* Banach space, and let $\mathcal{O} \subseteq X$ be a non-empty open set. We say that a function $f : \mathcal{O} \to \mathbb{C}$ is *holomorphic* if it is Fréchet differentiable in the usual (real) sense and if its Fréchet derivative is $\mathbb{C}$-linear at each point of $\mathcal{O}$. More precisely, f is holomorphic if for each $x \in \mathcal{O}$ there exists a bounded linear functional $L_x : X \to \mathbb{C}$ such that, for all $v \in X$,

$$L_x(v) \; = \; \lim_{\mathbb{C} \ni t \to 0} \frac{f(x + tv) - f(x)}{t} \; .$$

We write $L_x = Df(x)$, as usual.

The theory of holomorphic functions on several complex variables (say on $X = \mathbb{C}^n$, $n > 1$) is quite different from its one-dimensional counterpart $(X = \mathbb{C})$. Fortunately, however, several basic facts remain the same even if X is infinite dimensional. One of the most basic is the following.

Lemma 5.1.1 *If a sequence of holomorphic functions $f_n : \mathcal{O} \to \mathbb{C}$ converges uniformly to a (continuous) function $f : \mathcal{O} \to \mathbb{C}$ then f is holomorphic.* □

That this should be true in general is perhaps as surprising as it is in fact natural. Surprising, because compact sets in infinite-dimensional spaces seem too "thin" to hold much information; natural, if we take into account that holomorphic functions are *globally* determined as soon as they are known *locally*.

Another basic fact that we shall need is the following version of Schwarz's lemma, sometimes referred to as *Cauchy's inequality* (see e.g. [L6]).

Lemma 5.1.2 *Let $B \subset X$ be a ball of radius $R > 0$ and let $f : B \to \mathbb{D}$ be holomorphic. Then for all $x \in B$ we have*

$$\|Df(x)\| \le \frac{1}{R - \|x\|} \ . \tag{5.1}$$

Proof Let $u \in X$ be a unit vector and let $v = (R - \|x\|)u$. Define a map $\varphi_{x,u} : \mathbb{D} \to \mathbb{D}$ by

$$\varphi_{x,u}(t) = \frac{f(x + tv) - f(x)}{1 - \overline{f(x)}f(x + tv)} \ .$$

Then $\varphi_{x,u}(0) = 0$ and $\varphi_{x,u}$ is holomorphic (in t). Hence, by the classical Schwarz lemma, $|\varphi'_{x,u}(0)| \le 1$. But an easy computation yields

$$\varphi'_{x,u}(0) = \frac{Df(x)v}{1 - |f(x)|^2} \ .$$

Therefore $|Df(x)v| \le 1$ also, and this is equivalent to $|Df(x)u| \le 1/(R - \|x\|)$. Since u is an arbitrary vector of norm unity, we deduce that (5.1) holds true as asserted. □

This result allows us to prove the following version of Montel's theorem for holomorphic functions in Banach spaces. The concept of a normal family is the same as before. By lemma 5.1.1, all limits of sequences of elements of a normal family are holomorphic also.

Theorem 5.1.3 *Let $\mathcal{F}$ be a family of holomorphic functions $f : \mathcal{O} \to \mathbb{C}$. If $\mathcal{F}$ omits two values $a, b \in \mathbb{C}$, that is, if $f(\mathcal{O}) \subseteq \mathbb{C} \backslash \{a, b\}$ for all $f \in \mathcal{F}$, then $\mathcal{F}$ is a normal family.*

Proof Since normality is a local property, we may assume that $\mathcal{O} = B$, where B is a ball, say, of radius $R > 0$. Let $\pi : \mathbb{D} \to \mathbb{C} \backslash \{a, b\}$ be a holomorphic covering map (see theorem 3.5.1). Since B is simply connected, each map $f : B \to \mathbb{C} \backslash \{a, b\}$ in $\mathcal{F}$ lifts to a map $\hat{f} : B \to \mathbb{D}$, and $\hat{f}$ is holomorphic because π is locally bi-holomorphic. Consider now the family $\hat{\mathcal{F}} = \{\hat{f} : f \in \mathcal{F}\}$, and for each $r < R$ let $B_r \subset B$ be the *closed* ball of radius r concentric with B. We claim that $\hat{\mathcal{F}}$ is a normal family in each B_r. This will follow from the Arzelá–Ascoli theorem if we can prove both the following statements concerning $\hat{\mathcal{F}}$.

(a) The family $\hat{\mathcal{F}}$ is uniformly bounded in B_r. This is obvious, as $|\hat{f}(x)| < 1$ for all $x \in B$.

(b) The family $\hat{\mathcal{F}}$ is equicontinuous in B_r. It is in fact uniformly Lipschitz there because by lemma 5.1.2 we have $\|D\hat{f}(x)\| \leq 1/(R-r)$ for all $x \in B_r$ and so, by the mean-value inequality, $|\hat{f}(x) - \hat{f}(y)| \leq \|x - y\|/(R - r)$ for all $x, y \in B_r$ and each $\hat{f} \in \hat{\mathcal{F}}$.

This proves the claim, and it follows that $\hat{\mathcal{F}}$ is normal in B. But $\mathcal{F} = \{\pi \circ \hat{f} : \hat{f} \in \hat{\mathcal{F}}\}$ and π is continuous, whence $\mathcal{F}$ is normal in B as well. $\qquad\square$

One can of course go even further and state yet another version of Montel's theorem, the exact analogue of theorem 3.5.5. But the above version is sufficient for many applications.

We will to end this section with yet another version of Schwarz's lemma, which will be very useful later in the study of holomorphic motions. This new version requires us to consider holomorphic maps *into* Banach spaces, a generalization of the concept of holomorphic functions on Banach spaces. Let X, Y be complex Banach spaces, let $\mathcal{O} \subseteq X$ be an open set as before and let $f : \mathcal{O} \to Y$ be a map. We say that f is holomorphic if the composition of f with each holomorphic function defined on an open subset of Y containing $f(\mathcal{O})$ is holomorphic. With this formulation it is clear that compositions of holomorphic maps are holomorphic.

Now let B be an open ball in a complex Banach space X. We define a pseudo-distance $\mathrm{dist}_B(x, y)$ in the following way. Let $\mathcal{U}_{x,y}(\mathbb{D}, B)$ denote the set of maps $\varphi : \mathbb{D} \to B$ that are *univalent*, that is to say holomorphic and injective, and whose images $\varphi(\mathbb{D})$ contain both x and y. If $\|x - y\|$

is small compared with the Banach distances from both points to the boundary of B then the map $t \mapsto x + 2t(y - x)$ is an example of an element of $\mathcal{U}_{x,y}(\mathbb{D}, B)$. The reader may wonder whether $\mathcal{U}_{x,y}(\mathbb{D}, B)$ is always non-empty. We need not worry about that. We simply define functions $d_B^n : B \times B \to \mathbb{R}^+$, $n = 1, 2, \ldots$, writing first

$$d_B^1(x, y) = \inf_{\varphi \in \mathcal{U}_{x,y}(\mathbb{D}, B)} \text{dist}_{\mathbb{D}}(\varphi^{-1}(x), \varphi^{-1}(y)) \,,$$

where $\text{dist}_{\mathbb{D}}$ is the hyperbolic metric of the unit disk, and then setting

$$d_B^n(x, y) = \inf \sum_{i=0}^{n-1} d_B^1(x_i, x_{i+1})$$

where the infimum is taken over all chains of points $x_0, x_1, \ldots, x_n \in B$ with $x_0 = x$ and $x_n = y$. These functions satisfy $d_B^{n+1}(x, y) \leq d_B^n(x, y)$ for all n. Hence the limit

$$\text{dist}_B(x, y) = \lim_{n \to \infty} d_B^n(x, y)$$

exists for all $x, y \in B$. This defines a pseudo-distance, usually called the *Kobayashi* pseudo-distance of B. By an abuse of language, we shall refer to it below as the hyperbolic distance of B. Note that this pseudo-distance has the following property.

Lemma 5.1.4 *If $B = B(0, 1)$ is the unit ball of a complex Banach space then for each $x \in B$ we have*

$$\text{dist}_B(0, x) \leq \log \frac{1 + \|x\|}{1 - \|x\|} \,.$$

Proof We may assume that $x \neq 0$, otherwise there is nothing to prove. Let $\varphi : \mathbb{D} \to B$ be the map given by

$$\varphi(t) = \frac{tx}{\|x\|} \,.$$

This is a holomorphic injection with $\varphi(0) = 0$ and $\varphi(\|x\|) = x$. Hence we have

$$\text{dist}_B(0, x) \leq \text{dist}_{\mathbb{D}}(\varphi^{-1}(0), \varphi^{-1}(x)) = \text{dist}_{\mathbb{D}}(0, \|x\|) = \log \frac{1 + \|x\|}{1 - \|x\|} \,,$$

as asserted. $\qquad\square$

Our last version of Schwarz's lemma, by no means the most general we could state, is the following.

Lemma 5.1.5 *Let B be an open ball in a complex Banach space, let $V \subseteq \widehat{\mathbb{C}}$ be a hyperbolic domain and let $f : B \to V$ be a holomorphic map. Then f is a contraction of the corresponding hyperbolic distances, in other words, $\text{dist}_V(f(x), f(y)) \leq \text{dist}_B(x, y)$ for all $x, y \in B$.*

Proof First we claim that the desired inequality holds when d_B^1 replaces dist_B on the right-hand side. Indeed, given $\varepsilon > 0$, let $\varphi \in \mathcal{U}_{x,y}(\mathbb{D}, B)$ be such that

$$\text{dist}_{\mathbb{D}}(\varphi^{-1}(x), \varphi^{-1}(y)) < d_B^1(x, y) + \varepsilon \ .$$

Since $f \circ \varphi : \mathbb{D} \to V$ is holomorphic, the classical Schwarz lemma tells us that $\text{dist}_V(f(x), f(y)) \leq \text{dist}_{\mathbb{D}}(\varphi^{-1}(x), \varphi^{-1}(y))$. Therefore we have

$$\text{dist}_V(f(x), f(y)) \leq d_B^1(x, y) + \varepsilon \,,$$

and since ε is arbitrary the claim follows. Now, if we take any sequence $x_0, x_1, \ldots, x_n$ with $x_0 = x$ and $x_n = y$ then

$$
\begin{aligned}
\text{dist}_V(f(x), f(y)) &\leq \sum_{i=0}^{n-1} \text{dist}_V(f(x_i), f(x_{i+1})) \\
&\leq \sum_{i=0}^{n-1} d_B^1(x_i, x_{i+1}) = d_B^n(x, y) \ .
\end{aligned}
$$

Since this holds for all n, the lemma is proved. $\qquad\square$

This lemma shows in particular that the Kobayashi pseudo-distance dist_B is a distance after all! To show that $\text{dist}_B(x, y) > 0$ when x and y are distinct, all one has to do is find a holomorphic map, $f : B \to \mathbb{D}$, say, with $f(x) \neq f(y)$. There are plenty of such linear (affine) maps; simply use the Hahn–Banach theorem.

With a little extra effort, one can prove that the Kobayashi distance and the Banach distance are comparable away from the boundary of B. More precisely, if $B = B(0, R)$ then for all $0 < r < R$ there exists a constant $C(r) > 1$ with $C(r) \to \infty$ as $r \to R$ such that for all $x, y \in B(0, r)$ we have

$$\frac{1}{C(r)} \|x - y\| \leq \text{dist}_B(x, y) \leq C(r) \|x - y\| \ .$$

The proof is left as an exercise.

The reader is warned that there is a far-reaching general theory of complex hyperbolic spaces. We have only scratched the surface here. See [La] for more.

5.2 Extension and quasiconformality of holomorphic motions

Intuitively at least, a holomorphic motion of a set $E \subseteq \widehat{\mathbb{C}}$ is an isotopy of E whose time parameter changes holomorphically in some open ball B of a complex Banach space. The general concept of holomorphic motion can be precisely formulated as follows.

Definition 5.2.1 *Let E be a subset of the Riemann sphere having at least three points. Let B be a ball centered at a point λ_0 in some complex Banach space. A holomorphic motion of E over B is a mapping*

$$f : B \times E \to \widehat{\mathbb{C}}$$

with the following properties:

(i) *for each $\lambda \in B$, the map $f_\lambda : E \to \widehat{\mathbb{C}}$ given by $f_\lambda(z) = f(\lambda, z)$ is injective;*

(ii) *the map f_{λ_0} is the identity;*

(iii) *for each $z \in E$, the map $\lambda \mapsto f_\lambda(z)$ is holomorphic.*

Geometrically, a holomorphic motion is a collection of codimension-1 holomorphic submanifolds of the product $B \times \widehat{\mathbb{C}}$ that are pairwise disjoint, transversal to the fibers of the projection onto the first factor and such that the intersection of the union of these submanifolds with the fiber over λ_0 is $\{\lambda_0\} \times E$.

The Ahlfors–Bers theorem gives a family of examples of holomorphic motions. Indeed, if μ_λ is a holomorphic family of Beltrami coefficients with L^∞ norms uniformly bounded away from 1 and vanishing at $\lambda = 0$ then the corresponding family f_λ of quasiconformal homeomorphisms fixing $0, 1, \infty$ is a holomorphic motion of the whole Riemann sphere. Note that this family of examples has an extra property that is not present in the definition: continuity on both variables. The remarkable fact that we will discuss in this section is that this continuity is a consequence of the other conditions, because any holomorphic motion over the unit disk in the complex plane is the restriction of one of these holomorphic motions of the sphere given by the Ahlfors–Bers theorem.

5.2.1 The λ-lemma

The first important result on the extension and quasiconformality of holomorphic motions is the so called λ-*lemma* of Mañé, Sad and Sullivan [MSS], which guarantees the extension of a holomorphic motion of any set to a holomorphic motion of the closure of that set and also the continuity of the motion in both variables. Already from this extension result many dynamical consequences follow, such as the density of structural stability in the space of all polynomials of degree d. The λ-lemma also states that each map $f_\lambda : E \to \widehat{\mathbb{C}}$ is quasiconformal in the following sense.

Definition 5.2.2 *An injective map $f : E \to \widehat{\mathbb{C}}$ is said to be quasiconformal if there exists a continuous, increasing and surjective function $\sigma : [0, \infty] \to [0, \infty]$ such that*

$$|[f(z_1), f(z_2), f(z_3), f(z_4)]| \leq \sigma(|[z_1, z_2, z_3, z_4]|)$$

for any four points $z_1, z_2, z_3, z_4 \in E$.

Here $[z_1, z_2, z_3, z_4] = (z_1 - z_2)(z_3 - z_4)/(z_1 - z_3)(z_2 - z_4)$ denotes the cross-ratio of the four points. If E is a subdomain of $\widehat{\mathbb{C}}$, this definition of quasiconformality implies the previous one, given in Chapter 4; when $E = \widehat{\mathbb{C}}$ the two definitions are equivalent, a fact established by J. Väisälä.

Theorem 5.2.3 (The λ-lemma) *Let $f : B \times E \to \widehat{\mathbb{C}}$ be a holomorphic motion of a set $E \subseteq \widehat{\mathbb{C}}$. Then the map $f_\lambda : \overline{E} \to \widehat{\mathbb{C}}$ given by $f_\lambda(z) = f(\lambda, z)$ is quasiconformal for each $\lambda \in B$. Moreover, f has an extension to a continuous map $\widehat{f} : B \times \overline{E} \to \widehat{\mathbb{C}}$ that is a holomorphic motion of the closure of E, and such an extension is unique.*

Proof There is no loss of generality if we think of B as the unit ball $B(0, 1)$. We assume that E has at least four points, otherwise there is nothing to prove. We may assume also that the points $0, 1, \infty$ belong to E and remain fixed throughout the motion (normalizing the motion using a suitable holomorphic family of Möbius transformations).

Let us first prove that for each $\lambda \in B$ the map $f_\lambda : E \to \widehat{\mathbb{C}}$ is quasiconformal. To do this we need to control the distortion of cross-ratios and it is obviously sufficient to control the distortion of *small* cross-ratios, i.e. those whose absolute value is not greater than 1. Hence, take four distinct points $z_1, z_2, z_3, z_4 \in E$ with $|[z_1, z_2, z_3, z_4]| \leq 1$ and let $g : B \to \mathbb{C}^{**} = \mathbb{C} \setminus \{0, 1\}$ be the map given by

$$g(\lambda) = [f_\lambda(z_1), f_\lambda(z_2), f_\lambda(z_3), f_\lambda(z_4)] .$$

Since the four points on the right-hand side remain pairwise distinct for all λ (because f_λ is injective), their cross-ratio is never equal to 0, 1 or ∞. Therefore g is a well-defined map, and it is also holomorphic.

Applying lemma 5.1.5 with $V = \mathbb{C}^{**}$ and then using lemma 5.1.4, we get

$$\text{dist}_{\mathbb{C}^{**}}(g(0), g(\lambda)) \leq \text{dist}_B(0, \lambda) \leq \log \frac{1 + \|\lambda\|}{1 - \|\lambda\|} . \qquad (5.2)$$

Now we need a good lower bound for the distance on the first left-hand side. At this point we will suppose that both $g(0)$ and $g(\lambda)$ belong to the unit disk. There are other cases to consider, and the reader is invited to fill in the blanks. Now, using the fact that the hyperbolic metric of $\mathbb{C}^{**}$ near zero is bounded from below by $|dz|/(|z| \log (|Az|^{-1}))$ for some constant $0 < A < 1$, it is possible to prove that

$$\text{dist}_{\mathbb{C}^{**}}(z, w) \geq \left| \log \log \frac{1}{|Az|} - \log \log \frac{1}{|Aw|} \right| , \qquad (5.3)$$

for all $z, w \in \mathbb{D}^*$. Combining (5.2) with (5.3) we deduce that

$$\frac{1 - \|\lambda\|}{1 + \|\lambda\|} \leq \frac{\log |Ag(\lambda)|}{\log |Ag(0)|} \leq \frac{1 + \|\lambda\|}{1 - \|\lambda\|} ,$$

and therefore, for some constant $C_\lambda > 0$, we have

$$\left| [f_\lambda(z_1), f_\lambda(z_2), f_\lambda(z_3), f_\lambda(z_4)] \right| \leq C_\lambda \left| [z_1, z_2, z_3, z_4] \right|^{(1 - \|\lambda\|)/(1 + \|\lambda\|)} . \qquad (5.4)$$

This shows that f_λ is quasiconformal for all $\lambda \in B$. It may also be shown after some work that each f_λ is Hölder continuous with exponent $1/K_\lambda$, where $K_\lambda = (1 + \|\lambda\|)/(1 - \|\lambda\|)$, with respect to the *spherical metric*. In other words, one can prove from (5.4) that

$$\text{dist}_{\widehat{\mathbb{C}}}(f_\lambda(z), f_\lambda(w)) \leq M_\lambda \, \text{dist}_{\widehat{\mathbb{C}}}(z, w)^{1/K_\lambda} \qquad (5.5)$$

for all $z, w \in E$, for some $M_\lambda > 0$; once again the proof is left as an exercise.

We are now in a position to show that $f : B \times E \to \widehat{\mathbb{C}}$ extends uniquely to $B \times \overline{E}$ as a holomorphic motion. For each $z \in E$, let $\phi_z : B \to \mathbb{C}^{**}$ be the map given by $\phi_z(\lambda) = f(\lambda, z)$. Given $w \in \overline{E} \setminus E$, let $z_n \in E$ be a sequence converging to w. We already know from theorem 5.1.3 that $\{\phi_{z_n}\}$ is a normal family, so that any limit will be holomorphic, but we claim that in fact this sequence *converges* uniformly on compact subsets of B to a (holomorphic) function $\phi_w : B \to \mathbb{C}^{**}$. Indeed, from (5.5) we have

$$\text{dist}_{\widehat{\mathbb{C}}}(\phi_{z_m}(\lambda), \phi_{z_n}(\lambda)) \leq M_\lambda \, \text{dist}_{\widehat{\mathbb{C}}}(z_m, z_n)^{1/K_\lambda} ,$$

so that $\{\phi_{z_n}(\lambda)\}$ is a Cauchy sequence in the spherical metric for each $\lambda \in B$. Therefore the limit $\phi_w(\lambda) = \lim_{n \to \infty} \phi_{z_n}(\lambda)$ exists, and the same argument shows that it is independent of the choice of sequence z_n in E converging to w. Hence we have a well-defined map $\phi_w : B \to \mathbb{C}^{**}$ that is necessarily holomorphic. Thus we define $\widehat{f} : B \times \overline{E} \to \widehat{\mathbb{C}}$, taking $\widehat{f}(\lambda, z) = f(\lambda, z)$ if $z \in E$ and $\widehat{f}(\lambda, w) = \phi_w(\lambda)$ if $w \in \overline{E} \setminus E$. This extends the motion in an obviously unique way to a holomorphic motion of $\overline{E}$. $\qquad \square$

5.2.2 A compactness principle for holomorphic motions

We note the following consequence of the proof of theorem 5.2.3. Roughly speaking, the result states that a holomorphic motion is Lipschitz continuous in the first variable and Hölder continuous in the second, with respect to the spherical metric $d_{\widehat{\mathbb{C}}}$.

Proposition 5.2.4 *For each $0 < r < 1$, there exist positive constants a, b, α, depending only on r, such that the following holds. If $f : B_1 \times E \to \widehat{\mathbb{C}}$ is a holomorphic motion then*

$$d_{\widehat{\mathbb{C}}}(f(\lambda_1, z_1), f(\lambda_2, z_2)) \leq a|\lambda_1 - \lambda_2| + b d_{\widehat{\mathbb{C}}}(z_1, z_2)^\alpha .$$

Proof This is left as an exercise for the reader. $\qquad \square$

This fact, in turn, gives rise to the following useful compactness principle for holomorphic motions.

Corollary 5.2.5 *Let $\{0, 1, \infty\} \subseteq E_1 \subseteq E_2 \subseteq \cdots$ be an increasing sequence of subsets of the Riemann sphere, and let $E = \cup_{n \geq 1} E_n$. Suppose that $f_n : B_1 \times E_n \to \widehat{\mathbb{C}}$, $n = 1, 2, \ldots$, is a sequence of normalized holomorphic motions. Then there exist a normalized holomorphic motion $f : B_1 \times \overline{E} \to \widehat{\mathbb{C}}$ and a sequence $n_k \to \infty$ such that f_{n_k} converges to f uniformly in $B_r \times E_n$ for each $r < 1$ and each n.*

Proof Use the Arzelá–Ascoli theorem: clearly the sequence (f_n) is uniformly bounded in the spherical metric and proposition 5.2.4 above tells us that it is equicontinuous also, in each set $B_r \times E_n$. The details are again left as an exercise. $\qquad \square$

5.2.3 Further results

A remarkable improvement of the λ-lemma was obtained by Sullivan and Thurston in [ST]. They proved that a holomorphic motion of any set with at least four points over a ball of radius 1, when restricted to a ball whose radius is a universal positive number, can be extended to a holomorphic motion of the whole Riemann sphere and, in particular, that the maps f_λ become quasiconformal homeomorphisms of the entire Riemann sphere. This result was improved by Bers and Royden in [BR], see section 5.3 below. They proved that we can add a regularity condition onto the extension and obtain not only the existence but also the uniqueness of the extension. Both results hold for holomorphic motions over a ball on any complex Banach space. But in general it is not possible to extend the initial motion: it is necessary first to make a restriction to motion over a ball of radius one-third of the original ball. Thus, the philosophy behind these results is "restrict in time and extend in space". However, if the motion is over a simply connected domain in the complex plane then the extension is possible without time restriction, by the theorem below.

Theorem 5.2.6 (Extension of holomorphic motions) *Let $\Lambda \subset \mathbb{C}$ be a simply connected domain and E be a subset of the Riemann sphere with at least four points. Let $h : \Lambda \times E \to \widehat{\mathbb{C}}$ be a holomorphic motion with basepoint $\lambda_0 \in \Lambda$. Then h admits a continuous extension $H : \Lambda \times \widehat{\mathbb{C}} \to \widehat{\mathbb{C}}$ having the following properties:*

(i) *H is a holomorphic motion of $\widehat{\mathbb{C}}$;*
(ii) *$H_\lambda : \widehat{\mathbb{C}} \to \widehat{\mathbb{C}}$ is K-quasiconformal with K bounded by the hyperbolic distance between λ and the basepoint λ_0.*

This theorem was proved by Slodkowski in [Sl].

5.3 The Bers–Royden theorem

Let us now move to the statement of the theorem of Bers and Royden mentioned above. We need to introduce an important concept. Let V be a domain in the Riemann sphere that is covered by the disk, and let $\pi : \mathbb{D} \to V$ be the holomorphic universal covering map. A *quadratic differential* on V is represented by a function $\varphi : \mathbb{D} \to \mathbb{C}$ such that

$$\varphi(\gamma(z))(\gamma'(z))^2 = \varphi(z)$$

for all elements γ of the automorphism group of π and for every $z \in \mathbb{D}$. We recall that a Beltrami differential on V is represented by a function

$\mu : \mathbb{D} \to \mathbb{C}$ such that

$$\mu(\gamma(z))\frac{\overline{\gamma'(z)}}{\gamma'(z)} = \mu(z)$$

for every $z \in \mathbb{D}$ and for every element γ of the automorphism group of π. Also, a conformal metric in V is represented by a function $\rho : \mathbb{D} \to \mathbb{C}$ such that

$$\rho(\gamma(z))|\gamma'(z)| = \rho(z).$$

Therefore, $\mu = \overline{\varphi}/(\rho)^2$ represents a Beltrami differential on V whenever φ represents a quadratic differential and ρ represents a conformal metric.

Definition 5.3.1 (Harmonic Beltrami differentials) *A Beltrami differential on a domain V is harmonic if its lift to the universal cover is of the form $\mu = \overline{\varphi}/(\rho)^2$, where φ is the lift of a holomorphic quadratic differential and $\rho(z)|dz|$ is the hyperbolic metric of $\mathbb{D}$.*

When studying holomorphic motions, the crucial fact to remember about harmonic Beltrami differentials is that they enjoy the following uniqueness property.

Lemma 5.3.2 *Let V be a domain on the Riemann sphere, and suppose that $h_1 : V \to h_1(V)$ and $h_2 : V \to h_2(V)$ are quasiconformal maps with harmonic Beltrami coefficients μ_1 and μ_2. If h_1 and h_2 are Teichmüller equivalent, in other words if there exists a conformal map $c : h_1(V) \to h_2(V)$ such that $h_2^{-1} \circ c \circ h_1 : V \to V$ is homotopic to the identity relative to the boundary of V, then $\mu_1 = \mu_2$.*

Proof This is left as an exercise (at least for those who are willing to consult the appendix). $\qquad\square$

These are all the elements we need to give a precise statement of the Bers–Royden theorem.

Theorem 5.3.3 (Bers–Royden) *Let $B = B(0, r)$ be a ball of radius r centered at the origin in some complex Banach space. Let $f : B \times E \to \widehat{\mathbb{C}}$ be a holomorphic motion of a subset E of the Riemann sphere having at least three points. Then the restriction of this holomorphic motion to the smaller ball $B(0, r/3)$ has a unique continuous extension $\widehat{f} : B(0, r/3) \times \widehat{\mathbb{C}} \to \widehat{\mathbb{C}}$ with the following properties.*

(i) *The map $\widehat{f}$ is a holomorphic motion of the whole Riemann sphere.*
(ii) *For each λ, $\widehat{f}_\lambda$ is quasiconformal.*

(iii) *The Beltrami coefficient μ_λ of $\widehat{f}_\lambda$ is a harmonic Beltrami differential in each connected component of $\widehat{\mathbb{C}} \setminus E$.*

(iv) *The mapping $\lambda \mapsto \mu_\lambda(z)$ is holomorphic for almost all z.*

The proof is rather long and will be given in the following section. For the original proof and more, see [BR].

5.3.1 Proof of the Bers–Royden theorem

The original proof of the Bers–Royden theorem, elegantly presented in [BR], is an outstanding example of how some beautiful ideas of Teichmüller theory can be put together to yield the solution of a difficult problem. For the basic definitions and relevant facts about Teichmüller spaces needed in this section, the reader should consult the appendix.

Let us recall the setup. We are given a holomorphic motion $f : B_1 \times E \to \widehat{\mathbb{C}}$ of a set $E \subset \widehat{\mathbb{C}}$, with time parameter in a ball B_1 of radius 1 in some complex Banach space. We seek an extension of this motion to a motion of the entire sphere, at the cost of reducing B_1 to a ball $B_{1/3}$ of radius $1/3$. There is no loss of generality in assuming from the beginning that the motion is normalized in such a way that $0, 1, \infty$ remain fixed. The proof is divided into three steps.

Step 1. Let us first assume that E is a finite set, say

$$E = \{0, 1, \infty, z_1, z_2, \ldots, z_n\}$$

with the z_i pairwise distinct and also distinct from $0, 1, \infty$. Then we let f induce a map $F : B_1 \to M_n$ given by

$$F(\lambda) = (f(\lambda, z_1), f(\lambda, z_2), \ldots, f(\lambda, z_n)),$$

where $M_n = \{(\zeta_1, \zeta_2, \ldots, \zeta_n) \in (\mathbb{C}^{**})^n : \zeta_i \neq \zeta_j \ \forall \ i \neq j\}$. Note that M_n is an open subset of $\mathbb{C}^n$. The map F is clearly holomorphic. The complement of E in the Riemann sphere, $\widehat{\mathbb{C}} \setminus E$, is a *Riemann surface* (see the appendix for the basic facts about Riemann surfaces). Hence we can consider its *Teichmüller space* $\mathrm{Teich}(\widehat{\mathbb{C}} \setminus E)$ (again, see the appendix).

Moreover, since $\widehat{\mathbb{C}} \setminus E$ is in fact a *hyperbolic* Riemann surface, we know from the uniformization theorem (see the appendix) that $\widehat{\mathbb{C}} \setminus E = \mathbb{H}/\Gamma$, where $\Gamma \subset PSL(2, \mathbb{R})$ is a discrete torsion-free subgroup. By the *Bers embedding theorem* (theorem A.2.5), $\mathrm{Teich}(\widehat{\mathbb{C}} \setminus E)$ embeds holomorphically into the space $B(\Gamma)$ of holomorphic quadratic differentials φ in the lower half-plane $\mathbb{H}^*$ that are Γ-invariant, in the sense that

$$\varphi(z) = \varphi(\gamma z)(\gamma'(z))^2 \qquad \text{for all } \gamma \in \Gamma ,$$

and which have a finite *Nehari norm*

$$\|\varphi\| = \sup_{x+iy\in\mathbb{H}^*} y^2|\varphi(x+iy)| .$$

The image of $\text{Teich}(\widehat{\mathbb{C}} \setminus E)$ under the Bers embedding is contained in the ball of radius $3/2$ about the origin in $B(\Gamma)$, with respect to the Nehari norm (this norm makes $B(\Gamma)$ into a complex Banach space; see the appendix).

Lemma 5.3.4 *There exists a holomorphic universal covering map*

$$p : \text{Teich}(\widehat{\mathbb{C}} \setminus E) \to M_n.$$

Proof Given $\tau \in \text{Teich}(\widehat{\mathbb{C}} \setminus E)$, let $\phi : \widehat{\mathbb{C}} \setminus E \to X$ be a quasiconformal homeomorphism, onto another Riemann surface X, representing τ. By the uniformization theorem, we may assume that $X \subset \widehat{\mathbb{C}}$ (and that $\widehat{\mathbb{C}} \setminus X$ has $n+3$ points). After further post-composition with a suitable Möbius transformation, we may assume also that ϕ fixes the points $0, 1, \infty$ (as limiting points), so that these three points belong to $\widehat{\mathbb{C}} \setminus X$. Let $p(\tau)$ be given by the remaining n points; in other words, define

$$p(\tau) = (\phi(z_1), \phi(z_2), \ldots, \phi(z_n)) \in M_n .$$

We proceed through the following steps.

(1) *The map p is well defined.* Let $\psi : \widehat{\mathbb{C}} \setminus E \to Y$ be another representative of τ, with $Y \subset \widehat{\mathbb{C}}$ containing $0, 1, \infty$ and ψ normalized to fix these three points (again as boundary points). The equivalence relation defining the Teichmüller space requires the existence of a conformal map $c : X \to Y$ such that $\psi^{-1} \circ c \circ \phi : \widehat{\mathbb{C}} \setminus E \to \widehat{\mathbb{C}} \setminus E$ homotopic to the identity relative to E. By Riemann's removable singularity theorem, c extends to a conformal map of the entire Riemann sphere, so it must be Möbius. But the homotopy condition entails that c must fix $0, 1$ and ∞, so c is the identity. Therefore $\psi(z_j) = c \circ \phi(z_j) = \phi(z_j)$ for all j, so p is well defined.

(2) *The map p is onto.* This is clear, since for any other set $E' \subset \widehat{\mathbb{C}}$ having exactly $n+3$ points, there exists an orientation-preserving diffeomorphism of the entire sphere mapping E onto E' (in any way we like), and such a diffeomorphism is obviously quasiconformal.

(3) *The map p is holomorphic.* Here we use the Bers embedding theorem, according to which there is a holomorphic embedding

$$\beta : \text{Teich}(\widehat{\mathbb{C}} \setminus E) \to B(\Gamma) \ .$$

By a corollary of the Riemann–Roch theorem, see theorem A.1.2, $B(\Gamma)$ is finite dimensional and its complex dimension is $3 \times 0 - 3 + (n + 3) = n$. Hence $B(\Gamma)$ is isomorphic to $\mathbb{C}^n$. In particular, $\text{Teich}(\widehat{\mathbb{C}} \setminus E)$ is an n-dimensional complex manifold (and β acts as a chart). Let $\mu \in L^\infty(\widehat{\mathbb{C}} \setminus E)$ represent an element of $\text{Teich}(\widehat{\mathbb{C}} \setminus E)$. We denote by f^μ the unique normalized quasiconformal (qc) homeomorphism of $\widehat{\mathbb{C}}$ whose Beltrami coefficient is equal to μ. Let $\varphi_1, \varphi_2, \ldots, \varphi_n \in B(\Gamma)$ be a *basis* of $B(\Gamma)$ with $\|\varphi_j\| < 1/2$ for all j. By the *Ahlfors–Weill section theorem* (theorem A.2.4), there exist (harmonic) Beltrami coefficients $\mu_1, \mu_2, \ldots, \mu_n \in L^\infty(\widehat{\mathbb{C}} \setminus E)$ such that $\beta(\mu_j) = \varphi_j$ for all j. Given $(t_1, t_2, \ldots, t_n) \in \mathbb{C}^n$ with each $|t_j|$ sufficiently small, $\eta = t_1\mu_1 + \cdots + t_n\mu_n$ is a Beltrami coefficient in $\widehat{\mathbb{C}} \setminus E$ (i.e. $\|\eta\|_\infty < 1$), and we can take the Beltrami coefficient ν whose normalized solution to the Beltrami equation on the sphere is $f^\nu = f^\eta \circ f^\mu$. A straightforward computation using the formula for the complex dilatation of a composition of quasiconformal maps yields

$$\nu = \frac{t_1\nu_1 + t_2\nu_2 + \cdots + t_n\nu_n + \mu}{1 - \overline{\mu}(t_1\nu_1 + t_2\nu_2 + \cdots + t_n\nu_n)} \ , \qquad (5.6)$$

where

$$\nu_j = \mu_j \circ f^\mu \frac{\overline{\partial f^\mu}}{\partial f^\mu} \qquad \text{for each } j \ .$$

Note from (5.6) that ν depends holomorphically on $t_1, t_2, \ldots, t_n$. Thus we have a holomorphic map from a small neighborhood of the origin in $\mathbb{C}^n$ onto a small neighborhood of $[\mu] \in \text{Teich}(\widehat{\mathbb{C}} \setminus E)$, given by

$$(t_1, t_2, \ldots, t_n) \mapsto [\nu] \in \text{Teich}(\widehat{\mathbb{C}} \setminus E) \ .$$

Applying the Ahlfors–Bers theorem, f^ν depends holomorphically on $(t_1, t_2, \ldots, t_n)$ so the same can be said of

$$p([\nu]) = (f^\nu(z_1), f^\nu(z_2), \ldots, f^\nu(z_n)) \in M_n \ .$$

This shows that p is holomorphic in a neighborhood of $[\mu] \in \text{Teich}(\widehat{\mathbb{C}} \setminus E)$, as claimed.

(4) *The map p is a covering map.* Let us consider the subgroup G of the modular group $\mathrm{Mod}(\mathbb{C} \setminus E)$ consisting of self-maps of the Teichmüller space $\mathrm{Teich}(\widehat{\mathbb{C}} \setminus E)$ of the form $[f^\mu] \mapsto g_*([f^\mu]) = [f^\mu \circ g^{-1}]$, where $g : \widehat{\mathbb{C}} \setminus E \to \widehat{\mathbb{C}} \setminus E$ is quasiconformal and fixes every point of E. It is known (see the appendix) that G acts properly discontinuously on $\mathrm{Teich}(\widehat{\mathbb{C}} \setminus E)$. Suppose that $[f^\mu] \in \mathrm{Teich}(\widehat{\mathbb{C}} \setminus E)$ is such that $g_*([f^\mu]) = [f^\mu]$. Then, by the definition of the Teichmüller equivalence relation, this means that $(f^\mu)^{-1} \circ f^\mu \circ g^{-1} = g^{-1}$ is homotopic to the identity in $\widehat{\mathbb{C}} \setminus E$, so the same is true of g. This shows that $g_* = \mathrm{id}$, and we have proved that G acts *freely* on $\mathrm{Teich}(\widehat{\mathbb{C}} \setminus E)$. Now suppose that we have two elements $[f^\mu]$ and $[f^\nu]$ of $\mathrm{Teich}(\widehat{\mathbb{C}} \setminus E)$ such that $p([f^\mu]) = p([f^\nu])$. This will happen if and only if $g = (f^\mu)^{-1} \circ f^\nu$ fixes every point of E (and so belongs to G) and therefore if and only if $[f^\mu]$ and $[f^\nu]$ are in the same G-orbit. This shows at once that $\mathrm{Teich}(\widehat{\mathbb{C}} \setminus E)/G \simeq M_n$ and that p is the quotient map and therefore also a covering map. Since $\mathrm{Teich}(\widehat{\mathbb{C}} \setminus E)$ is simply connected (it is homeomorphic to a ball, see the appendix), it follows that p is a universal covering map. $\qquad\square$

Now, let us go back to the map F that we defined at the beginning of the above proof. Since the unit ball B_1 is simply connected and $F : B_1 \to M_n$ is holomorphic, the lemma we have just proved implies that F lifts to a holomorphic map into Teichmüller space, i.e. there exists a holomorphic map $\widehat{F} : B_1 \to \mathrm{Teich}(\widehat{\mathbb{C}} \setminus E)$ such that $p \circ \widehat{F} = F$.

The composition of our map $\widehat{F}$ with the Bers embedding can still, by an abuse of notation, be denoted by $\widehat{F}$. So now $\widehat{F} : B_1 \to B(\Gamma)$ is holomorphic and $\widehat{F}(B_1)$ is contained in the ball of radius $3/2$ about the origin in $B(\Gamma)$. By the Schwarz lemma for holomorphic maps in Banach spaces, the ball $B_{1/3} \subset B_1$ is mapped into the ball of radius $1/2$ about the origin in $B(\Gamma)$. In other words, for each $\lambda \in B_{1/3}$, $\widehat{F}(\lambda) = \varphi_\lambda$ is a holomorphic quadratic differential in $\mathbb{H}^*$ that is Γ-invariant and satisfies $\|\varphi_\lambda\| < 1/2$. By the Ahlfors–Weill theorem A.2.4, the harmonic Beltrami differential $\widehat{\mu}_\lambda$ in the upper half-plane given by $\widehat{\mu}_\lambda(z) = -2y^2 \varphi_\lambda(\overline{z})$ for all $z \in \mathbb{H}$ has the following properties:

(1) $\widehat{\mu}_\lambda \in L^\infty(\mathbb{H})$ and $\|\widehat{\mu}_\lambda\|_\infty < 1$;
(2) $\widehat{\mu}_\lambda$ quotients down (via the universal covering map $\mathbb{H} \to \widehat{\mathbb{C}} \setminus E$) to a harmonic Beltrami differential $\mu_\lambda \in L^\infty(\widehat{\mathbb{C}} \setminus E) \equiv L^\infty(\widehat{\mathbb{C}})$ with $\|\mu_\lambda\|_\infty < 1$;

(3) μ_λ depends holomorphically on λ, for so does φ_λ;

(4) μ_λ represents an element of $\mathrm{Teich}(\widehat{\mathbb{C}} \setminus E)$ that is mapped to φ_λ by the Bers embedding.

As a consequence of (1)–(3) above, applying the Ahlfors–Bers theorem to the family $(\mu_\lambda)_{\lambda \in B_{1/3}}$ we get a corresponding family of normalized qc-homeomorphisms $f_\lambda : \widehat{\mathbb{C}} \to \widehat{\mathbb{C}}$ with $\overline{\partial} f_\lambda = \mu_\lambda \partial f_\lambda$ that varies holomorphically with λ (note in particular that f_0 is the identity). Now, let us write

$$f_\lambda(\widehat{\mathbb{C}} \setminus E) \;=\; \widehat{\mathbb{C}} \setminus \{0, 1, \infty, \zeta_{1,\lambda}, \zeta_{2,\lambda}, \ldots, \zeta_{n,\lambda}\} \,,$$

where $\zeta_{j,\lambda} = f_\lambda(z_j)$ for each $j = 1, 2, \ldots, n$. Then from (4) above and the fact that $p \circ \widehat{F} = F$, we see that $\zeta_{j,\lambda} = f(\lambda, z_j)$ for all j. Therefore, letting $\widehat{f} : B_{1/3} \times \widehat{\mathbb{C}} \to \widehat{\mathbb{C}}$ be given by $\widehat{f}(\lambda, z) = f_\lambda(z)$, we deduce that $\widehat{f}$ is a normalized holomorphic motion (check!) and, since $\widehat{f}(\lambda, z_j) = f_\lambda(z_j) = f(\lambda, z_j)$ for all j, that $\widehat{f}$ is an extension of $f|_{B_{1/3} \times E}$ to the whole Riemann sphere. This establishes the existence of an extended (harmonic) motion when E is a finite set.

Step 2. In order to remove the extra hypothesis that E is finite, we use an approximation argument. Let $E \subset \widehat{\mathbb{C}}$ be an arbitrary set, and consider an increasing sequence of finite subsets of E whose union is *dense* in E, say

$$\{0, 1, \infty\} \subseteq E_1 \subseteq E_2 \subseteq \cdots \subseteq E_n \subseteq E \,, \quad \bigcup_{n=1}^{\infty} E_n \text{ is dense in } E \,.$$

We may in fact assume that E is closed, by the Mañé–Sad–Sullivan theorem 5.2.3. By step 1 (the finite case), for each $n \geq 1$ there exists a normalized holomorphic motion $f_n : B_{1/3} \times \widehat{\mathbb{C}} \to \widehat{\mathbb{C}}$ that extends $f|_{B_{1/3} \times E_n}$ and is harmonic in $\widehat{\mathbb{C}} \setminus E_n$. From the compactness principle for holomorphic motions (lemma 5.2.5) we may also assume, passing to a subsequence if necessary, that (f_n) converges uniformly on compact subsets as $n \to \infty$, to a holomorphic motion $\widehat{f} : B_{1/3} \times \widehat{\mathbb{C}} \to \widehat{\mathbb{C}}$ that extends $f|_{B_{1/3} \times E}$. Therefore, all we have to do is to show that the extended motion is harmonic. Let $V \subseteq \widehat{\mathbb{C}} \setminus E$ be a connected component of the complement of E, and let ρ_V be its hyperbolic density. Also, let ρ_n denote the hyperbolic density of $\widehat{\mathbb{C}} \setminus E_n$. Since $E_n \subseteq E_{n+1} \subseteq E$ for all $n \geq 1$, the monotonicity of Poincaré metrics (implied by, say, lemma 3.3.1) tells us that $\rho_n(z) \leq \rho_{n+1}(z) \leq \rho_V(z)$ for all $z \in V$. This shows that $\rho_\infty(z) = \lim_{n \to \infty} \rho_n(z)$ exists pointwise in V and that

$\rho_\infty(z) \leq \rho_V(z)$ for all $z \in V$. Moreover, since each Poincaré metric $\rho_n(z) |dz|$ has Gaussian curvature -1, we know from exercise 3.2 that ρ_n satisfies the elliptic PDE

$$\frac{\partial^2}{\partial x^2}(\log \rho_n) + \frac{\partial^2}{\partial y^2}(\log \rho_n) \;=\; \rho_n^2 \,. \tag{5.7}$$

From this fact and standard estimates on this elliptic PDE (which we omit), it follows that (ρ_n) converges *uniformly* to ρ_∞ on compact subsets of V, and we also have

$$\frac{\partial^2}{\partial x^2}(\log \rho_n) \to \frac{\partial^2}{\partial x^2}(\log \rho_\infty) \qquad \text{and} \qquad \frac{\partial^2}{\partial y^2}(\log \rho_n) \to \frac{\partial^2}{\partial y^2}(\log \rho_\infty)$$

uniformly on compact subsets as $n \to \infty$. This shows that ρ_∞ also satisfies (5.7), so the corresponding metric $\rho_\infty(z) |dz|$ has constant negative curvature equal to -1.

Hence, to show that $\rho_\infty = \rho_V$ it suffices to prove that the conformal metric $\rho_\infty |dz|$ is *complete*. Let γ be a rectifiable curve in V joining a point of V to a point of its boundary, say $w \in \partial V$. We assume here that w is not one of the points $0, 1, \infty$; if it is then the argument to follow has to be slightly modified, and this is left as an exercise for the reader. We need to show that

$$\int_\gamma \rho_\infty(z)|dz| \;=\; +\infty \,. \tag{5.8}$$

To prove (5.8), let us consider a sequence (ζ_n) with $\zeta_n \in E_n$ such that $\zeta_n \to w$ as $n \to \infty$. Denoting by $\sigma(z,\zeta)|dz|$ the Poincaré metric of $\widehat{\mathbb{C}} \setminus \{0, 1, \infty, \zeta\}$, we obviously have

$$\int_\gamma \sigma(z,w)|dz| \;=\; +\infty$$

as well as

$$\lim_{n\to\infty} \int_\gamma \sigma(z,\zeta_n)|dz| \;=\; +\infty \,, \tag{5.9}$$

because $\sigma(z,\zeta)$ varies continuously with ζ. But $\sigma(z,\zeta_n) \leq \rho_m(z)$ for all $m > n$ by the monotonicity of Poincaré metrics and this implies, of course, that $\rho_\infty(z) \geq \sigma(z,\zeta_n)$ for all $n \geq 1$. Using this fact in (5.9), we get (5.8) as desired. We have thus proved that $\rho_\infty = \rho_V$.

Now let $\mu_n(\lambda, z)$ be the Beltrami coefficient of $f_n(\lambda, z)$ in $\widehat{\mathbb{C}} \setminus E_n$. Since μ_n is harmonic and depends holomorphically on λ, there exists $\psi_n(\lambda, z)$, holomorphic in z and anti-holomorphic in $\lambda \in B_{1/3}$, such that

$$\mu_n(\lambda, z) \;=\; \rho_n^{-2}(z)\, \overline{\psi_n(\lambda, z)} \,.$$

We have already proved that $\rho_n \to \rho_\infty = \rho_V$ uniformly on compact subsets of V. We also know from Chapter 3 that the hyperbolic density $\rho_n(z)$ at $z \in \widehat{\mathbb{C}} \setminus E_n$ is bounded above by $2/\text{dist}(z, E_n)$. Using this, the fact that $\|\mu_n\|_\infty < 1$ and again the monotonicity of Poincaré metrics, we get

$$|\psi_n(\lambda, z)| \leq |\mu_n(\lambda, z)| \left(\frac{2}{\text{dist}(z, E_n)} \right)^2$$
$$\leq \left(\frac{2}{\text{dist}(z, V)} \right)^2 .$$

This shows that ψ_n is uniformly bounded on compact subsets of $B_{1/3} \times V$. Hence, passing to a subsequence if necessary, we deduce that ψ_n converges uniformly on compact subsets to $\psi : B_{1/3} \times V \to \widehat{\mathbb{C}}$, a function holomorphic in z and anti-holomorphic in λ.

This shows that, after passing to a subsequence, $\mu_n(\lambda, z)$ converges uniformly on compact subsets of $B_{1/3} \times V$ to

$$\mu(\lambda, z) = \rho_V^{-2}(z) \overline{\psi(\lambda, z)} .$$

But this μ must be the Beltrami coefficient of $\widehat{f}(\lambda, z)$, and we have just shown that it is harmonic, as required. This finishes the existence part of the proof of the Bers–Royden theorem.

Step 3. It remains to prove uniqueness. Here we merely sketch the argument, sending the reader to [BR] for details. Let $\widehat{f}_1$ and $\widehat{f}_2$ be two holomorphic motions of the Riemann sphere over $B_{1/3}$, both extending $f|_{B_{1/3} \times E}$ and both harmonic in $\widehat{\mathbb{C}} \setminus E$. We want to show that $\widehat{f}_1 = \widehat{f}_2$. First we claim that the images of any connected component V of $\widehat{\mathbb{C}} \setminus E$ under these two motions agree at all times, in other words that $\widehat{f}_1(\lambda, V) = \widehat{f}_2(\lambda, V)$ for each $\lambda \in B_{1/3}$. Note that this equality holds true at $\lambda = 0$, because $\widehat{f}_1(0, \cdot) = \widehat{f}_2(0, \cdot) = \text{id}$. Also, since

$$\widehat{f}_1|_{B_{1/3} \times E} = \widehat{f}_2|_{B_{1/3} \times E} ,$$

it follows that for all $\lambda \in B_{1/3}$ we have $\widehat{f}_1(\lambda, V) = \widehat{f}_2(\lambda, V_\lambda)$, where V_λ is *some* connected component of $\widehat{\mathbb{C}} \setminus \widehat{f}_2(\lambda, E) = \widehat{\mathbb{C}} \setminus \widehat{f}_1(\lambda, E)$. From these facts and an easy continuity argument, we deduce that the set $A \subseteq B_{1/3}$ consisting of those λ for which $V_\lambda = V$ is both open and closed and is non-empty. Therefore $A = B_{1/3}$ because the latter set is connected, and this proves our claim. Thus, for each component V as above and each $\lambda \in B_{1/3}$ the map $g\lambda : V \to V$ given by $g_\lambda = \widehat{f}_2(\lambda, \cdot)^{-1} \circ \widehat{f}_1(\lambda, \cdot)$

is well-defined, quasiconformal and fixes every point of ∂V. Now, it can be proved that the quasiconformal maps $\widehat{f}_1(\lambda, \cdot)|_V$ and $\widehat{f}_2(\lambda, \cdot)|_V$ are Teichmüller equivalent, in the sense of lemma 5.3.2. This non-trivial fact depends on a result on quasiconformal isotopies and will not be proved here; see [BR, lemma II, p. 176]. Applying lemma 5.3.2, we see that the Beltrami coefficients of $\widehat{f}_1(\lambda, \cdot)|_V$ and $\widehat{f}_2(\lambda, \cdot)|_V$, both being harmonic, must coincide. But this means that g_λ is holomorphic and, since it fixes every point at the boundary of V, it must be the identity. This shows that the two motions agree in every connected component of $\widehat{\mathbb{C}} \setminus E$ at all times, and since they also agree in E (being extensions of the same f), it follows that $\widehat{f}_1 = \widehat{f}_2$ as desired. This establishes the uniqueness part of the Bers–Royden theorem.

5.4 Density of structural stability

Let us now move on to some interesting applications of these ideas to complex dynamics. We will show in this section how the theory of holomorphic motions can be used to investigate the structural stability properties of rational maps.

Let Λ be some open set in a finite-dimensional complex vector space (or even a complex manifold). A holomorphic family of rational maps of degree d is a holomorphic function $F : \Lambda \times \widehat{\mathbb{C}} \to \widehat{\mathbb{C}}$ such that, for each $\lambda \in \Lambda$, the map $F_\lambda : \widehat{\mathbb{C}} \to \widehat{\mathbb{C}}$ given by $F_\lambda(z) = F(\lambda, z)$ has topological degree d. Given such a family we can consider the set of *structurally stable* (*quasiconformally stable*) parameter values Λ_{ss} (Λ_{qcs}), which is the set of points $\lambda_0 \in \Lambda$ having a neighborhood $\mathcal{N} \subset \Lambda$ such that each F_λ for $\lambda \in \mathcal{N}$ is topologically conjugate (quasiconformally conjugate) to F_λ. Note that Λ_{ss} and $\Lambda_{qcs} \subset \Lambda_{ss}$ are open subsets of Λ. The number of stable cycles of a map in Λ_{ss} is a locally constant function of the map. Hence Λ_{ss} is contained in the open set Λ_{Js} of J-*stable* parameter values, which is precisely the set of parameter values where the number of attracting cycles is locally constant.

Lemma 5.4.1 *Let* $F : \Lambda \times \widehat{\mathbb{C}} \to \widehat{\mathbb{C}}$ *be a holomorphic family of rational maps of degree* $d \geq 2$. *Let* z_0 *be an indifferent fixed point of* F_{λ_0}. *If there are sequences* $\lambda_n \to \lambda_0$, $z_n \to z_0$ *such that* $f_{\lambda_n}(z_n) = z_n$ *and* $|F'_{\lambda_n}(z_n)| > 1$ *then there are sequences* $\hat{\lambda}_n \to \lambda_0$ *and* $\hat{z}_n \to z_0$ *such that* $F_{\hat{\lambda}_n}(\hat{z}_n) = \hat{z}_n$ *and* $|F'_{\hat{\lambda}_n}(\hat{z}_n))| < 1$.

Proof Suppose first that $F'_{\lambda_0}(z_0) \neq 1$. Then, by the implicit function theorem, there are neighborhoods $\mathcal{N}$ of λ_0 and V of z_0 and a holomorphic function $\lambda \in \mathcal{N} \mapsto z(\lambda) \in V$ such that $z(\lambda)$ is the unique fixed point of F_λ in V. The function $\lambda \mapsto F_\lambda'(z(\lambda))$ is holomorphic, and it is not constant since there are repelling fixed points accumulating at z_0. Hence its image covers a full neighborhood of 1, which proves the existence of attracting fixed points converging to z_0. Next, suppose that $F_{\lambda_0}'(z_0) = 1$. Let B be a small ball centered at z_0 such that F_{λ_0} does not have fixed points in the boundary of B and such that $|F_{\lambda_0}'(w) - 1| < 1/2$ for all $w \in B$. Let $\mathcal{N}$ be a neighborhood of λ_0 such that the same properties hold for F_λ in this neighborhood. For each $\lambda \in \mathcal{N}$ the function F_λ has $k = k(\lambda)$ fixed points $z_1(\lambda), z_2(\lambda), \dots, z_k(\lambda)$ in B. We claim that the function $d(\lambda) = \prod_{j=1}^{k} F_\lambda'(z_j(\lambda))$ is holomorphic. In fact, let Log be a branch of the logarithmic function on the ball of radius 1 centered at 1. The poles of the meromorphic function on B defined by

$$\frac{(F_\lambda'(z) - 1) \operatorname{Log} F_\lambda'(z)}{F_\lambda(z) - z}$$

are precisely the fixed points of F_λ whose derivative is different from unity, and the residue of this meromorphic function at each fixed point is the logarithm of the derivative of F_λ at that fixed point. Therefore, by the residue theorem, we have

$$d(\lambda) = \exp\left(\frac{1}{2\pi i} \int_{\partial B} \frac{(F_\lambda'(z) - 1) \operatorname{Log} F_\lambda'(z)}{F_\lambda(z) - z} \, dz \right),$$

and this is clearly a holomorphic function of λ. The existence of repelling fixed points implies that this function is not constant. Hence its image contains a full neighborhood of 1, which once again implies the existence of attracting fixed points. □

Theorem 5.4.2 *Let* $F \colon \Lambda \times \widehat{\mathbb{C}} \to \widehat{\mathbb{C}}$ *be a holomorphic family of rational maps of degree* $d \geq 2$. *Then:*

(i) *the set of J-stable parameter values* Λ_{Js} *is open and dense in* Λ;

(ii) *if* $\lambda_0 \in \Lambda_{Js}$, *there exists a ball* $B \subset \Lambda_{Js}$ *centered at* λ_0 *and a holomorphic motion* h_λ *of the Riemann sphere over* B *such that* h_λ *maps the Julia set of* F_{λ_0} *onto the Julia set of* F_λ *and conjugates the two maps on their Julia sets.*

Proof By definition, Λ_{Js} is an open set. The number of attracting periodic cycles of F_λ is a lower semicontinuous function that is uniformly bounded. Hence, this function is locally constant in a neighborhood of any local maximum, which implies the density of Λ_{Js}. This proves (i).

In order to prove (ii), let $B \subset \Lambda_{Js}$ be a ball centered at λ_0. By the implicit function theorem, the set of hyperbolic attracting fixed points moves holomorphically over this ball B. In particular the eigenvalues at the attracting periodic points are bounded away from the unit circle. If $p(\lambda_0)$ is a repelling periodic point of F_{λ_0} of period N then, again by the implicit function theorem, there exists a holomorphic function $p(\lambda)$ in a neighborhood of λ_0 such that $p(\lambda)$ is a repelling periodic point of F_λ of the same period N. This function $p(\lambda)$ extends holomorphically to the ball B because otherwise there would be a sequence $\lambda_n \to \lambda \in B$ such that the repelling fixed point $p(\lambda_n)$ of $F_{\lambda_n}^N$ converged to an indifferent fixed point $p(\lambda)$ of F_λ^N. By lemma 5.4.1, there would exist parameter values close to λ for which the Nth iterate of the corresponding map has an attracting fixed point with eigenvalue arbitrarily close to the unit circle, but this is impossible.

Now, if $q(\lambda_0)$ is another repelling periodic point of F_{λ_0} of period N distinct from $p(\lambda_0)$ then $q(\lambda) \neq p(\lambda)$ for all λ because otherwise there would exist λ such that $p(\lambda) = q(\lambda)$ is an indifferent fixed point of F_λ^{MN}, which is not possible, again by lemma 5.4.1. Thus we have a holomorphic motion of the set of repelling periodic points of F_{λ_0} that respects the dynamics, i.e. conjugates F_{λ_0} to F_λ on the set of repelling periodic points. By the Bers–Royden theorem, this motion extends to a holomorphic motion h_λ of the whole Riemann sphere (provided that we restrict the motion first to a ball concentric with B with one-third of its radius), and since the repelling periodic points are dense in the Julia set, the restriction of h_λ to $J(F_{\lambda_0})$ conjugates F_{λ_0} with the restriction of F_λ to the h_λ-image of the Julia set of F_λ. Since the Julia set of F_{λ_0} is totally invariant, it follows that its h_λ-image is totally invariant under F_λ and therefore coincides with the Julia set of F_λ. This proves (ii), and we are done. $\qquad\square$

Lemma 5.4.3 *Let $F \colon \Lambda \times \widehat{\mathbb{C}} \to \widehat{\mathbb{C}}$ be a holomorphic family of rational maps of degree $d \geq 2$. Then there exists an open and dense subset $\Lambda_0 \subset \Lambda$ such that for each $\lambda_0 \in \Lambda_0$ there is a neighborhood $\mathcal{N} \subset \Lambda_0$ and holomorphic functions $c_1(\lambda), \ldots, c_k(\lambda)$ on $\mathcal{N}$ such that the set of critical points of F_λ is precisely $\{c_1(\lambda), \ldots, c_k(\lambda)\}$.*

Proof The sum of the multiplicities of the critical points does not depend on λ and is a bound for the total number of critical points. Hence the set Λ_0 of local maxima for the number of critical points is dense. Let $\lambda_0 \in \Lambda_0$ and let $c_j(\lambda_0), j = 1, \ldots, k$, be the critical points of F_{λ_0}. Denote by m_j

the multiplicity of $c_j(\lambda_0)$. Let $B_j, j = 1, \ldots, k$, be pairwise disjoint small balls centered at $c_j(\lambda_0)$. For λ near λ_0, the sum of the multiplicities of the critical points of F_λ in each B_j is equal to m_j. Since the total number of critical points is equal to k, F_λ has a unique critical point $c_j(\lambda) \in B_j$ of multiplicity m_j. It remains to prove that $c_j(\lambda)$ is holomorphic in λ. But the residue theorem gives us the formula

$$m_j c_j(\lambda) = \frac{1}{2\pi i} \int_{\partial B_j} z \, \frac{F''_\lambda(z)}{F'_\lambda(z)} \, dz \ ,$$

from which it is clear that $c_j(\lambda)$ is a holomorphic function. $\square$

The lemma below was proved in [Av].

Lemma 5.4.4 *For $j = 1, 2$, let h^j_λ be holomorphic motions of the subsets X_j of the Riemann sphere with the same basepoint. Suppose that X_1 has an empty interior and $h^2_\lambda(x) \in h^1_\lambda(X_1)$ if and only if $x \in X_1 \cap X_2$. Then $h^2_\lambda(x) = h^1_\lambda(x)$ for every $x \in X_1 \cap X_2$ and there is a holomorphic motion of the union $X_1 \cup X_2$ that is an extension of both motions.*

Proof We may suppose that B is the unit disk in the complex plane and that the basepoint is 0. Let $\hat{h}^1_\lambda \colon \widehat{\mathbb{C}} \to \widehat{\mathbb{C}}$ be an extension of the holomorphic motion h^1_λ to a holomorphic motion of the Riemann sphere. Let $Y = \{x\} \cup (\widehat{\mathbb{C}} \setminus X_1)$. Clearly, the mapping $\mathbb{D} \times Y \to \widehat{\mathbb{C}}$ defined by $x \mapsto h^2_\lambda(x)$ and $y \mapsto \hat{h}^1_\lambda(y)$, $y \in \widehat{\mathbb{C}} \setminus X_1$, is a holomorphic motion and therefore has an extension to a holomorphic motion of the Riemnan sphere. Since X_1 has empty interior this extension must coincide with $\hat{h}^1_\lambda$ by the continuity of $\hat{h}^1_\lambda$. $\square$

Lemma 5.4.5 *Let $\lambda_0 \in \Lambda_0 \cap \Lambda_{\mathrm{Js}}$. Then there exist a ball $B \subset \Lambda_0 \cap \Lambda_{\mathrm{Js}}$ centered at λ_0 and a holomorphic motion $h_\lambda \colon \widehat{\mathbb{C}} \to \widehat{\mathbb{C}}$ with the following properties:*

(i) *h_λ is a conjugacy between the Julia sets, the sets of attracting periodic points, the sets of Siegel periodic points and the sets of super-attracting periodic points;*

(ii) *h_λ preserves critical sets and the multiplicities of the critical points.*

(iii) *h_λ preserves the indifferent periodic points and their type;*

(iv) *h_λ maps Fatou components onto Fatou components of the same type.*

Proof Since $\lambda_0 \in \Lambda_{Js}$, there is a holomorphic motion of the Julia set over a ball centered at λ_0. By the previous lemma, if we take this ball to be small enough then there are holomorphic functions $c_1(\lambda), \ldots, c_k(\lambda)$ on this ball that yield the critical points of the maps F_λ. If $c_j(\lambda_0) \in J(F_{\lambda_0})$ then the restriction of F_{λ_0} to a neighborhood of $c_j(\lambda_0)$ in the Julia set is $m_j - 1$ to 1 and, since h_λ is a conjugacy at the Julia set, the same is true for $h_\lambda(c_j(\lambda_0))$. Therefore $h_\lambda(c_j(\lambda_0)) = c_j(\lambda)$. Thus we can extend the motion of the Julia set to a motion of the union of the Julia set and the critical set, defining it to be $c_i(\lambda)$ if $c_i(\lambda_0)$ does not belong to the Julia set. Since the set of periodic attractors is disjoint from the set of critical points and also from the Julia set, we can extend the holomorphic motion, shrinking B if necessary, to a holomorphically varying conjugacy in the set of periodic attractors. Since the eigenvalues at the periodic attractors of F_λ are bounded away from zero and bounded away from the unit circle, it follows that if $c_j(\lambda_0)$ is periodic of period N then it remains periodic of the same period because otherwise we could find an attracting periodic point of some F_λ with eigenvalue as close to zero as we wished. Similarly, if $p(\lambda_0)$ is an indifferent periodic point of period N in the Julia set then the derivative of $F_\lambda^N(h_\lambda(p_\lambda(0)))$, which is a holomorphic function of λ, must be constant. Otherwise, for some λ the point $h_\lambda(p(\lambda_0))$ would be an attractor, which is not possible. If the indifferent point is not in the Julia set then it is the center of a Siegel disk and has a periodic analytic continuation that must have the same derivative, since there are no attractors with small eigenvalues. Therefore we can extend the holomorphic motion to be a conjugacy also in the set of indifferent periodic points in the Fatou set. $\square$

The positive orbits of two critical points of a map in the family may have an intersection that might be destroyed even by an arbitrarily small perturbation of the parameter. Also, if the intersection of both critical orbits is empty, a small perturbation may create an intersection. This is obviously an obstruction to the structural stability of the map. We are going to prove that this is the only obstruction and that the set of parameter values where this obstruction is not present is open and dense in the parameter space. A parameter value $\lambda_0 \in \Lambda_0$ is *post-critically stable* if the following condition is satisfied: if $c_1(\lambda), \ldots, c_k(\lambda)$ are the critical points of F_λ, for λ in a small neighborhood of λ_0, and if $F_\lambda^a(c_i(\lambda)) = F_\lambda^b(c_j(\lambda))$ for some λ and some $a, b \in \mathbb{N}$ then the equality is an identity in a neighborhood of λ_0. The set of post-critically stable parameters will be denoted by Λ_{Ps}.

Lemma 5.4.6 *The set of post-critically stable parameter values Λ_{Ps} is an open and dense subset of the parameter space.*

Proof For each $\lambda \in \Lambda_0 \cap \Lambda_{\mathrm{Js}}$ let $X_0(\lambda)$ be the union of the Julia set with the set of attracting and super-attracting periodic points and the set of periodic points of Siegel type. Let $X(\lambda)$ be the union of $X_0(\lambda)$ with the post-critical set of F_λ. By lemma 5.4.5, $X_0(\lambda)$ moves holomorphically in the neighborhood of each point in $\Lambda_0 \cap \Lambda_{\mathrm{Js}}$ through conjugacies. Since $\Lambda_0 \cap \Lambda_{\mathrm{Js}}$ is dense it is enough to prove that this holomorphic motion extends to a holomorphic motion by conjugacies of $X(\lambda)$ based at each point of a dense subset. Let $\lambda_0 \in \Lambda_0 \cap \Lambda_{\mathrm{Js}}$ and B_0 be a ball centered at λ_0. We have to prove the existence of post-critically stable parameter values in B_0. Taking B_0 small enough we have that $X_0(\lambda)$ moves holomorphically over B_0 through conjugacies and extends to a holomorphic motion that includes the critical set $\{c_1(\lambda), \ldots, c_k(\lambda)\}$. Let $X_j(\lambda)$ be the union of $X_0(\lambda)$ with the closure of the orbits of the critical points $c_1(\lambda), \ldots, c_j(\lambda)$ together with all the leaves of the dynamical laminations of the super-attracting basins that intersect these forward orbits. We will prove by induction that for each j there exist $\lambda_j \in B_{j-1}$, a ball $B_j \subset B_{j-1}$ centered at λ_j and a holomorphic motion by conjugacies of X_j over B_j with basepoint λ_j. Suppose, by induction, that we have constructed the holomorphic motion of X_{j-1}. Let us consider the critical point $c_j(\lambda_{j-1})$. We will have to deal with several possibilities.

(1) There is a neighborhood $\mathcal{N}$ of λ_{j-1} and an integer a such that $F_\lambda^a(c_j(\lambda)) \in X_{j-1}(\lambda)$ for all $\lambda \in \mathcal{N}$. Let a be the smallest integer satisfying this property. Taking a small ball B_j centered at λ_{j-1} and contained in $\mathcal{N}$, we may assume also that $F_\lambda^j(c_j(\lambda)) \in X_{j-1}(\lambda)$ for every $\lambda \in B_j$. If $c_j(\lambda)$ belongs to the Julia set we may take $\lambda_j = \lambda_{j-1}$ and $B_j = B_{j-1}$, since the original motion is already a conjugacy in the Julia set. So suppose that $c_j(\lambda)$ belongs to the Fatou set and that the forward orbit of the critical point $c_j(\lambda_{j-1})$ moves holomorphically. Since $X_{j-1}(\lambda_{j-1})$ has empty interior (otherwise it would be the whole Riemann sphere and the lemma would be proved already), there is a holomorphic motion that extends both motions by lemma 5.4.4.

(2) The point $c_j(\lambda_{j-1})$ is in the basin of an attracting or parabolic component. Here we have two sub-cases.

(2a) The forward orbit of $c_j(\lambda_{j-1})$ is disjoint from the forward orbits of the critical points $c_1(\lambda_{j-1}), \ldots, c_{j-1}(\lambda_{j-1})$. Then the same disjointness property holds for λ in a neighborhood of λ_{j-1}. Thus the holomorphic motion of the critical orbit of c_j is disjoint from the holomorphic motion of X_{j-1} and both define a holomorphic motion of the union of X_{j-1} with the forward orbit of c_j. By the λ-lemma, the motion of this union extends to its closure, which is X_j.

(2b) If (2a) does not hold then either (1) holds or we can approximate λ_{j-1} by a λ_j for which (2a) holds.

(3) There is an iterate $F_{\lambda_{j-1}}^a(c_j)$ that belongs to a Siegel disk or to a Herman ring. Again we have two sub-cases.

(3a) The leaf of the foliation through $F_{\lambda_{j-1}}^a(c_j)$ does not belong to X_{j-1}. The argument here is similar to the one in (2a): the holomorphic motion of the forward orbit of c_j in a small ball B_j centered at λ_{j-1} is disjoint from the motion of X_{j-1} and so we arrive at the same conclusion.

(3b) If (3a) does not hold then either (1) holds or we can approximate λ_{j_1} by a λ_j for which (3a) holds.

(4) There is an iterate $F_{\lambda_{j-1}}^a(c_j)$ that falls in a circle leaf of a super-attracting basin. Again we have to deal with two sub-cases.

(4a) None of the forward orbits of the critical points c_i with $i < j$ intersects the circle leaf through $F_{\lambda_{j-1}}^a(c_j(\lambda_{j-1}))$. Here we extend the motion using the same argument as in (3a).

(4b) If (4a) does not hold then either (1) holds or we can approximate λ_{j-1} by a λ_j for which (4a) holds.

Since we have considered all the possibilities for the dynamical behavior of the forward orbit of c_j the lemma is proved. $\qquad\square$

Theorem 5.4.7 *The post-critically stable parameter values coincide with the qc-stable parameter values.*

Proof Clearly, the set of qc-stable parameter values is contained in the set of post-critically stable parameter values. It remains to prove the opposite inclusion.

(1) If λ_0 is a post-critical stable parameter value then there exists a ball B_0 centered at λ_0 and a holomorphic motion of the post-critical set over B_0 with basepoint λ_0.

(2) The holomorphic motion in (1) can be extended to a holomorphic motion of the grand orbits of the critical points, preserving the dynamics. To see why, let $\Phi\colon B_0 \times \widehat{\mathbb{C}} \to B_0 \times \widehat{\mathbb{C}}$ be the mapping $\Phi(\lambda, z) = (\lambda, F_\lambda(z))$.

Let $C_0 = \{(\lambda, c_j(\lambda)) \in B_0 \times \widehat{\mathbb{C}}; j = 1, \ldots, k\}$. By (1), $\cup_{n=0}^{\infty}(\Phi^n(C_0))$ is a disjoint union of holomorphic codimension-1 sub-manifolds transversal to the fibers of the projection in the first coordinate, i.e. a disjoint union of graphs of holomorphic functions from B^0 into $\widehat{\mathbb{C}}$. In particular, $C_1 = \Phi(C_0)$ is a finite union of such graphs. By the implicit function theorem, $\Phi^{-1}(C_1)$ is also the union of C_0 with a finite number of graphs of holomorphic functions. The restriction of Φ gives a degree-d holomorphic covering map of $(B_0 \times \widehat{\mathbb{C}}) \setminus \Phi^{-1}(C_1)$ onto $(B_0 \times \widehat{\mathbb{C}}) \setminus C_1$, and the pre-image of a family of disjoint holomorphic graphs by this covering is a family of disjoint holomorphic graphs. Hence, iterating backwards the components of $\cup_{n=0}^{\infty}(\Phi^n(C_0))$ we obtain a family of disjoint graphs of holomorphic functions, yielding a holomorphic motion of the grand orbit of the critical points of F_{λ_0}.

(3) When restricted to a ball of radius equal to one-third of the radius of B_0, the holomorphic motion in (2) has a unique extension to a harmonic holomorphic motion of the Riemann sphere *by conjugacies*. Indeed, by the λ-lemma, the motion constructed in (2) extends to the holomorphic motion h_λ of the closure of the grand orbits $\hat{P}(\lambda)$ of the critical orbits. By the Bers–Royden theorem, the restriction of this motion over the ball B, of radius $1/3$ of the radius of B_0, can be extended uniquely to a holomorphic motion h_λ of the Riemann sphere whose Beltrami differential is harmonic in each component of $\widehat{\mathbb{C}} \setminus \hat{P}(\lambda)$. For each $\lambda \in B$, the restriction

$$F_\lambda : \widehat{\mathbb{C}} \setminus \hat{P}(\lambda) \to \widehat{\mathbb{C}} \setminus \hat{P}(\lambda)$$

is a covering map and we can lift h_λ to a map $\hat{h}_\lambda$ such that $h_{\lambda_0} \circ F_{\lambda_0} = F_\lambda \circ \hat{h}_\lambda$. Furthermore, $\hat{h}_\lambda$ coincides with $\hat{h}_\lambda$ in $\hat{P}(\lambda_0)$, where h_λ is a conjugacy. Since the the restriction of F_λ maps each connected component of $\widehat{\mathbb{C}} \setminus \hat{P}(\lambda)$ holomorphically onto another component it is a local isometry of the respective hyperbolic metrics. It follows that the Beltrami coefficient of $\hat{h}_\lambda$ is also harmonic. By uniqueness $\hat{h}_\lambda = h_\lambda$, proving that h_λ is a conjugacy. This shows that λ_0 is a qc-stable parameter, which completes the proof. $\square$

Corollary 5.4.8 *The set of quasiconformally stable parameter values is dense.* $\square$

Corollary 5.4.9 *The set of structurally stable parameter values is dense.* $\square$

This of course achieves the main goal of this section, namely, to show that structural stability is a dense property in $\mathrm{Rat}_d(\widehat{\mathbb{C}})$.

Remark 5.4.10 *In the family of all polynomial maps of degree d there are no Herman rings, and no Siegel disk can persist under all small perturbations. Hence the stable maps of this family do not have Siegel disks or Herman rings. The same is true for the family of all rational maps of degree d, but in this case the proof that no map with a Herman ring can be stable is much more involved. This result was obtained by Mañé in [Man].*

5.5 Yoccoz rigidity: puzzles and para-puzzles

The mathematical tools under discussion have been especially successful in describing the intricate dynamical behaviour and bifurcation structure of the family $\{f_c : z \mapsto z^2 + c\}_{c \in \mathbb{C}}$ of quadratic polynomials.

For each $c \in \mathbb{C}$, let K_c denote the filled-in Julia set of f_c, i.e. the set of points having bounded forward orbits. As we have already seen, if the critical point itself does not belong to K_c then K_c is a Cantor set. Otherwise K_c is connected, since it is the intersection of a nested family of topological disks $f_c^{-n}(\{z : |z| < R\})$. The *Mandelbrot set* $\mathcal{M}$ (see figure 5.1) is the set of parameter values $c \in \mathbb{C}$ such that the filled-in Julia set K_c is connected. Some of the most basic properties of the Mandelbrot set are summarized in the following classical result of Douady and Hubbard [DH1].

Theorem 5.5.1 (Douady–Hubbard) *The Mandelbrot set $\mathcal{M}$ is a compact and simply connected set contained in the closed disk of radius 2 about the origin in the complex plane.*

Proof Note that for all c and all $n \geq 0$ we have $|f_c^{n+1}(0)| \geq |f_c^n(0)|^2 - |c|$. Hence we get by induction

$$|f_c^n(0)| \geq |c| (|c| - 1)^{2^{n-1}} .$$

It follows that if $|c| > 2$ then $|f_c^n(0)| \to \infty$ as $n \to \infty$. This shows that $\mathcal{M} \subset \{|c| \leq 2\}$. Now we claim that $\mathcal{M}$ is precisely the set of those c values with $|c| \leq 2$ such that $|f_c^n(0)| \leq 2$ for all $n \geq 0$. Indeed, if for such c values and some $m > 0$ we have $|f_c^m(0)| = 2 + \delta$ with $\delta > 0$ then

$$|f_c^{m+1}(0)| \geq (2 + \delta)^2 - 2 = 2 + 4\delta .$$

Therefore by induction

$$|f_c^{m+k}(0)| \geq 2 + 4^k \delta$$

for all $k \geq 0$, so that $|f_c^{m+k}(0)| \to \infty$, proving that $c \notin \mathcal{M}$. This proves our claim, which shows that $\mathcal{M}$ is closed and therefore compact. Now look at $\widehat{\mathbb{C}} \setminus \mathcal{M}$. Suppose that this set had a bounded component $\mathcal{O}$. Then for all $c \in \partial\mathcal{O}$ we would have $|f_c^n(0)| \leq 2$, for all $n \geq 0$. But $f_c^n(0)$ is a holomorphic function of c (it is in fact a polynomial), and so by the maximum principle we would have $|f_c^n(0)| \leq 2$ for all $c \in \mathcal{O}$, for all $n \geq 0$, which would force $\mathcal{O} \subset \mathcal{M}$, a contradiction. Therefore $\widehat{\mathbb{C}} \setminus \mathcal{M}$ has only a component containing ∞, and this proves that $\mathcal{M}$ is simply connected as asserted. $\qquad\square$

Douady and Hubbard proved also that $\mathcal{M}$ is *connected*. See [DH1] for the original proof and the comments below.

Now let $c \in \mathcal{M}$. Since K_c is simply connected there exists by the Riemann mapping theorem a unique holomorphic diffeomorphism $\phi_c \colon \widehat{\mathbb{C}} \setminus K_c \to \widehat{\mathbb{C}} \setminus \overline{\mathbb{D}}$ that maps ∞ to ∞ and is normalized in such a way that its derivative at ∞ is a positive real number. Then $\phi_c \circ f_c \circ \phi_c^{-1}$ is a degree-2 holomorphic covering map from $\widehat{\mathbb{C}} \setminus \overline{\mathbb{D}}$ onto itself and maps each sequence converging to the boundary $\partial\mathbb{D}$ into a sequence that also converges to $\partial\mathbb{D}$. Hence, by the Schwarz reflection principle ([Rud, p. 260]) it extends to a degree-2 branched covering of the sphere, with ∞ and 0 as fixed branched points. It follows that $\phi_c \circ f_c \circ \phi_c^{-1} = f_0$. We will consider the two families of curves $R(\theta, c) = \phi_c^{-1}(\{re^{2\pi i\theta}; r > 1\})$ for $0 \leq \theta < 1$ and $E(\rho, c) = \phi_c^{-1}(\{z; |z| = e^\rho\})$ for $\rho > 0$. Each curve $R(\theta, c)$ is called an *external ray* of angle θ. The closed curves $E(\rho, c)$ are called *equipotentials*, since they are the level curves of the Green function of $\mathbb{C} \setminus K_c$, namely $G_c \colon \mathbb{C} \setminus K_c \to \mathbb{R}$, $G_c(z) = \log|\phi_c(z)|$. Since ϕ_c is a conjugacy, we have $f_c(R(\theta, c)) = R(2\theta, c)$, where the angles are measured mod 1, and $f_c(E(\rho, c)) = E(2\rho, c)$. As $\rho \to 0$, the equipotentials $E(\rho, c)$ converge to the Julia set J_c, which is the boundary of K_c. We say that the external ray $R(\theta, c)$ *lands* at a point $z \in J_c$ if $z = \lim_{r \to 1} \phi_c^{-1}(\{re^{2\pi i\theta}; r > 1\})$. A fundamental result for the description of the combinatorial behavior of quadratic polynomials is the following theorem, which is due to Douady and Hubbard and to Yoccoz.

Theorem 5.5.2 *If θ is rational then the external ray $R(\theta, c)$ lands at a point of the Julia set that is either periodic or eventually periodic.*

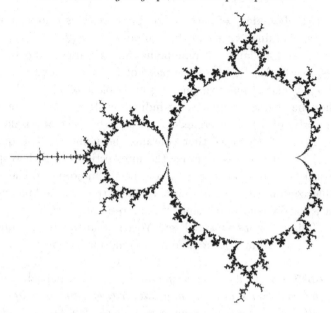

Fig. 5.1. The Mandelbrot set's boundary.

Each repelling or parabolic periodic point is the landing point of a finite number of periodic external rays.

The external ray of angle 0 lands at a fixed point denoted by β. If $c \neq 1/4$ is in the boundary of the Mandelbrot set then f_c has another fixed point α, which, by the above theorem, is the landing point of at least two external rays. We can now use these external rays to define a sequence of partitions of neighborhoods of K_c that have a Markov property. We start with a topological disk $\mathcal{P}_0$ bounded by an equipotential $E(\rho_0, c)$. The external rays that land at the fixed point α decompose this topological disk into a finite number of components, which are the *puzzle pieces* of the next partition P_1. Since each puzzle piece is bounded by pieces of the external rays landing at α and pieces of the equipotential $E(\rho_0, c)$ and since f_c permutes the external rays and maps the equipotential $E(\rho_0, c)$ onto the equipotential $E(2\rho_0, c)$ that surrounds $E(\rho_0, c)$, it follows that the Markov property holds for P_1; in other words, if the image of a puzzle piece intersects another puzzle piece then it contains this puzzle piece. Now we can define inductively the Markov partitions P_n as follows: the puzzle piece $P_n(z)$ that contains a point $z \in K_c$ which is not mapped into α under f_c^j, $j \leq n$, is the component of $f_c^{-1}(P_{n-1}(f_c(z)))$ that contains z.

Note that the boundary of each puzzle piece of P_n is a finite union of curves, each contained either in the equipotential $E(2^{-(n-1)}\rho_0, c)$ or in an external ray that lands in some point that is mapped to α by f_c^j for some $j \leq n - 1$. Also, each puzzle piece of P_n is contained in a puzzle piece of P_{n-1} and is mapped onto a puzzle piece of P_{n-1}. Therefore it satisfies the Markov property, by induction. If $z \in K_c$ is not in the backward orbit of α then there exists a nested sequence of puzzle pieces $P_0(z) \supset P_1(z) \supset P_2(z) \cdots$ that contains the point z. The question considered by Yoccoz was whether the intersection of all these puzzle pieces consists of the point z only; of special interest are the puzzle pieces that contain the critical point. If this is the case we find that the Julia set is locally connected, since the intersection of each puzzle piece with the Julia set is a connected set. Yoccoz's answer to this question requires the important notion of complex renormalization.

Definition 5.5.3 *A quadratic polynomial f is renormalizable if there exist a topological disk U containing the critical point and an integer $p \geq 2$, called the renormalization period, such that the restriction of f^p to U is a quadratic-like map $f^p \colon U \to V$ having a connected filled-in Julia set. Let K_p be the filled-in Julia set of the quadratic-like map $f^p \colon U \to V$. We say that f is simply renormalizable if, furthermore, the intersection of each pair of the sets $K_p, f(K_p), \ldots, f^{p-1}(K_p)$ is either empty or is a unique point that does not disconnect any of them.*

We may now state Yoccoz's theorem as follows.

Theorem 5.5.4 (Yoccoz) *If $c \in \partial \mathcal{M}$ is such that f_c is non-renormalizable and does not have an indifferent periodic point then the intersection of a nested sequence of puzzle pieces containing a point in the Julia set is this point. In particular the Julia set is locally connected.*

To prove this theorem Yoccoz considered the region $A_n(0)$ between the critical puzzle pieces $P_n(0)$ and $P_{n+1}(0)$. If the boundary of A_{n+1} is in the interior of $P_n(0)$ then $A_n(0)$ is a ring domain, and we can consider its modulus mod $A_n(0)$. Otherwise, i.e. if the boundary of $P_{n+1}(0)$ does touch the boundary of $P_n(0)$, we define the modulus of $A_n(0)$ to be equal to zero. The main estimate in the proof of Yoccoz's theorem entails that the series $\sum \text{mod } A_n(0)$ is divergent; see [Hu].

The above theorem can be modified to cover also the case where the maps are only finitely renormalizable. In this case we start the puzzle construction using the external rays that land at the periodic point of the polynomial that corresponds to the fixed point α of the last renormalization.

Since ∞ is a super-attracting fixed point of f_c, there is a holomorphic diffeomorphism ϕ_c of neighborhoods of ∞ (the so-called Böttcher coordinates) conjugating f_c with f_0 and normalized to be a tangent to the identity at ∞. The pre-images by ϕ_c of circles give an f_c-invariant foliation by analytic simple closed curves in a neighborhood of ∞. Taking the backward iterates of this foliation by f_c we get an f_c-invariant foliation of the complement of K_c, which is singular at the backward orbit of the critical value if $c \notin \mathcal{M}$. The leaf through the critical point is a figure-eight curve, and ϕ_c extends holomorphically to a conjugacy of the unbounded component of the complement of this leaf. This unbounded component contains the critical value c. Douady and Hubbard proved in [DH1] that the mapping $\Phi \colon c \in \mathbb{C} \setminus \mathcal{M} \mapsto \phi_c(c)$ is a holomorphic diffeomorphism onto $\mathbb{C} \setminus \overline{\mathbb{D}}$ (which shows, in particular, that $\mathcal{M}$ is connected; note that this fact does not follow from theorem 5.5.1). The set $\mathcal{R}(\theta) = \Phi^{-1}(\{Re^{2\pi i\theta}; R > 1\})$ is the *external ray* of angle θ in the parameter space, and the closed curve $\mathcal{E}(\rho) = \{c \in \mathbb{C} : |\Phi(c)| = e^\rho\}$ is also called an equipotential. This gives a foliation of the complement of the Mandelbrot set by analytic curves starting at ∞ and approaching the Mandelbrot set. If the external ray approaches a unique point in the Mandelbrot set then we say that it *lands* at that point, just as before. Now we have the following parameter space version of theorem 5.5.4.

Theorem 5.5.5 (Yoccoz) *If $c \in \mathcal{M}$ is such that f_c is not infinitely renormalizable then $\mathcal{M}$ is locally connected at c.*

The now famous MLC conjecture states that the boundary of the Mandelbrot set is locally connected. The above theorem of Yoccoz reduces the analysis of this conjecture to understanding the infinitely renormalizable points. It is known that the MLC conjecture implies the Fatou conjecture for the quadratic family. These topics are well beyond the scope of this book, and therefore we will merely refer the reader to [DH2] and [Hu].

Exercises

5.1   Let $f : \mathbb{D} \times S^1 \to \widehat{\mathbb{C}}$ be the map $f(\lambda, z) = z + \lambda z^{-1}$. Show that f is a holomorphic motion and find an explicit extension of f to a holomorphic motion of the entire Riemann sphere.

5.2   Let $g : \mathbb{D} \times S^1 \to \widehat{\mathbb{C}}$ be the holomorphic motion of the unit circle
given by $g(\lambda, z) = f(\lambda^2, z)$, where f is as in the previous exer-
cise. Show that every extension of g to a holomorphic motion
of the closed unit disk, $\hat{g} : \mathbb{D} \times \overline{\mathbb{D}} \to \widehat{\mathbb{C}}$, has the form

$$\hat{g}(\lambda, z) = z + h(z)\lambda + \overline{z}\lambda^2 \ ,$$

where h is a continuous real-valued function on the closed unit
disk that vanishes at the boundary.

6

The Schwarzian derivative and cross-ratio distortion

In this chapter we will describe the main tool in real one-dimensional dynamics: the distortion of the cross-ratio, and its relation to the Schwarzian derivative.

6.1 Cross-ratios and the Schwarzian

The dynamical systems we will consider are generated by smooth (C^r, $r \geq 3$) maps $f: X \to X$, where X is either the circle S^1 or a compact interval $[0, 1]$ of the real line. We will also assume that each critical point of f is non-flat, i.e. if $Df(p) = 0$ then $f(p+x) = f(p)+(\phi(x))^k$ for small x, where $k \geq 2$ is an integer and ϕ is a smooth local diffeomorphism with $\phi(0) = 0$. In particular f has only a finite number of critical points.

There are only three differential operators acting on smooth mappings of the real line that are relevant to dynamics in the sense that they behave well under composition of mappings. The first differential operator is just the derivative $f \mapsto Df = f'$ with the chain rule $D(f \circ g) = Df \circ g \cdot Dg$. The second is the *non-linearity*

$$ f \mapsto Nf = \frac{D^2 f}{Df} = D \log Df $$

whenever $Df \neq 0$. In this case, the chain rule reads

$$ N(f \circ g) = Nf \circ g \cdot Dg + Ng \ . $$

Finally, the third differential operator is the *Schwarzian derivative*, defined as

$$ Sf = \frac{D^3 f}{Df} - \frac{3}{2} \left(\frac{D^2 f}{Df} \right)^2 = D(Nf) - \frac{1}{2}(Nf)^2 \ . $$

155

In this case we have the chain rule

$$S(f \circ g) = Sf \circ g \cdot (Dg)^2 + Sg .$$

The kernel of the non-linearity operator N in the space of orientation-preserving smooth mappings is the group of affine transformations, whereas the kernel of the Schwarzian derivative is the group of Möbius transformations. From the last formula it follows that any composition of maps having negative Schwarzian derivatives also has a negative Schwarzian derivative. Hence every iterate f^n of a map with negative Schwarzian derivative also has a negative Schwarzian derivative. Another important consequence of this formula is that the Schwarzian derivative is negative in a small neighborhood of any non-flat critical point. What makes maps with negative Schwarzian derivative dynamically interesting is the fact that such maps expand the cross-ratio of four points. If M is an interval compactly contained in another interval T then the cross-ratio of the pair (T, M) is

$$\mathrm{Cr}(T, M) = \frac{|T||M|}{|L||R|} ,$$

where $|T|$ denotes the length of the interval T and R, L are the components of $T \setminus M$. If $\phi: T \to \mathbb{R}$ is a diffeomorphism onto its image, the distortion of the cross-ratio is denoted by

$$C(\phi; T, M) = \frac{\mathrm{Cr}(\phi(T), \phi(M))}{\mathrm{Cr}(T, M)} .$$

If ϕ has negative Schwarzian derivative then $C(\phi; T, M) > 1$. To see this, first note that composing ϕ with a linear map that reverses orientation (and preserves the cross-ratio), we may assume that ϕ preserves orientation. Also, composing ϕ with a Möbius transformation we may assume that ϕ fixes the endpoints of T and L. If the cross-ratio is not expanded we must have $\phi(M) \subset M$ and, therefore, $R \subset \phi(R)$. By the mean-value theorem there exist points $x_1 \in L$, $x_2 \in M$ and $x_3 \in R$ such that $D\phi(x_1) = 1$, $D\phi(x_2) < 1$ and $D\phi(x_3) > 1$. If follows that $D\phi$ has a minimum at a point $x \in [x_1, x_3]$. Therefore $D^2\phi(x) = 0$ and $D^3\phi(x) \geq 0$ and the Schwarzian derivative is non-negative at x. This contradiction shows that maps with negative Schwarzian derivative expand the cross-ratio.

From this simple fact we get an interesting dynamical consequence. Let p be an attracting fixed point of a mapping f and let B be its immediate basin of attraction, i.e. B is the maximal interval around p such that all iterates in B converge to p. We claim that B must contain a turning point of f. In fact, if this were not the case then f would be monotone in B and, considering f^2 instead of f if f is orientation-reversing, we could assume that the endpoints of B are fixed points of f. Let $M \subset B$ be the interval bounded by p and another point in B. Then f contracts the cross-ratio $C(B, M)$, since B is fixed by f, M is contracted, one component of $B \setminus M$ is fixed and the other is expanded. This contradiction shows that f must have a critical point in the immediate basin of attraction. This proves a theorem of Singer [Sin], according to which the number of periodic attractors of a mapping with negative Schwarzian derivative is bounded by the number of critical points.

The cross-ratio has also an interesting interpretation in terms of hyperbolic geometry. In fact, we can embed the interval T as a geodesic of the hyperbolic space modeled by a disk $D \subset \mathbb{C}$ having T as diameter. To compute the hyperbolic length of the middle interval M we consider the Möbius transformation ϕ that maps the left endpoint of T to 0, the right endpoint of L to i and the right endpoint of T to ∞. Hence the disk D is mapped into the upper half-space $\mathbb{H}$, the interval T into the vertical geodesic and the interval M into the interval bounded by i and mi, $m > 1$, that has the same hyperbolic length as M, which is therefore equal to $\log m$. Since a Möbius transformation preserves cross-ratios and the cross-ratio of the four points $0, i, mi, \infty$ is equal to $m - 1$, we have that the hyperbolic length of M is equal to $\log(1 + C(T, M))$. The reader should compare the expansion of the cross-ratio by maps with negative Schwarzian derivative with the fact that a critically finite rational map expands the hyperbolic metric of the complement of the post-critical set in the Riemann sphere (see Chapter 3).

To go beyond maps with negative Schwarzian derivative, the main tool introduced in [MvS] is control of the distortion of the cross-ratio of two nested intervals under diffeomorphic iteration. The crucial fact is that if f is a C^2 map with non-flat critical points then there exists a constant $\gamma(f) > 0$ such that if $M \subset T \subset N$ and the derivative of f does not vanish at T then

$$C(f; T, M) \geq 1 - \gamma(f)|T| .$$

From this estimate it follows that if $l > 0$ then there exists a positive constant C_l such that if $\sum_{i=0}^{n-1} |f^i(T)| \leq l$ and $f^n|T$ is a diffeomorphism then $C(f^n; T, M) > C_l$. The constant C_l is equal to unity for all l if f has negative Schwarzian derivative. In general it depends only on f

and converges to 1 as $l \to 0$. See section 6.2 below for an infinitesimal formulation of this distortion tool.

The relevance of this estimate is related to the *real Koebe distortion principle*, which can be stated as follows.

Lemma 6.1.1 *There exists a positive constant $D = D(\tau, C)$ such that, for any C^1 diffeomorphism $\phi \colon T \to \mathbf{R}$ with $C(\phi; T', M') > C$ for all $M' \subset T' \subset T$, which also satisfies*

$$\min\left\{ \frac{|\phi(L)|}{|\phi(M)|}, \frac{|\phi(R)|}{|\phi(M)|} \right\} > \tau\,,$$

the distortion of ϕ at the middle interval M, namely

$$D(h, M) = \sup_{x,y \in M} \frac{|Dh(x)|}{|Dh(y)|}\,,$$

is bounded by D. The constant D tends to 1 as $(\tau, C) \to (\infty, 1)$.

Proof See [MS, Chapter IV]. □

Hence we get good control of the non-linearity in the middle interval as long as the distortion of the cross-ratio is bounded from below and the image of the middle interval is well inside the image of the total interval. From this lemma it follows also that Koebe space can be pulled back to the domain: given τ and C as above, there exists τ' depending only on τ and C such that

$$\min\left\{ \frac{|L|}{|M|}, \frac{|R|}{|M|} \right\} > \tau'\,.$$

So, in order to get good control of the distortion of an iterate on an interval M, we have to embed M into a larger interval T in such a way that the iterate of f is still diffeomorphic on T, the total length of the iterates of the larger interval is uniformly bounded and we have definite Koebe space around the image of the middle interval. The bound for the total length is usually obtained by some weak disjointness property: we say that a family of intervals has intersection multiplicity bounded by k, or is k-quasidisjoint, if each point is contained in at most k intervals of the family. In most applications k is a fixed integer that depends on the number of critical points; see [SV]. To construct Koebe space, the typical procedure is to look at the interval of minimal length in a family of disjoint intervals, where we have space on both sides, and then use Koebe's distortion principle to pull back the space. This is illustrated in the case of circle maps in section 6.2.

A major result obtained using these ideas is the theorem that states the non-existence of wandering intervals for C^2 maps with non-flat critical points. A non-wandering interval is an interval I such that its forward orbit $f^j(I), j \in \mathbb{N}$, is a family of pairwise disjoint intervals and such that I is not in the basin of attraction of a periodic point. This theorem has a very long history. It was first proved by Denjoy [Den] for diffeomorphisms of the circle. In this case the distortion is controlled using only the non-linearity operator. Next it was proved by Yoccoz [Yo] for smooth circle homeomorphisms with non-flat critical points. In [Gu] it was proved for unimodal maps with negative Schwarzian derivative and in [MvS] for C^2 unimodal maps. The multimodal case without critical points of the inflexion type was proved in [L8] and [BL] and the final result was stated in [MMS]; see also [SV] for an easier proof. The same tools were used in [MMS] to prove that the periods of non-expanding periodic points are bounded. More precisely, there exist constants $\lambda(f) > 1$ and $n(f) \in \mathbb{N}$ such that if p is a periodic point of f of period $n > n(f)$ then $|Df^n(p)| > \lambda(f)$. In particular, if f is real analytic then the number of attracting periodic points is bounded.

Another important application of these tools is to the theory of renormalization. We say that a unimodal map f is renormalizable if there exists an interval I_0 around the critical point such that the first return map to I_0 is again a unimodal map whose domain is I_0. This means that the forward orbit of I_0 is a cycle of intervals with pairwise disjoint interiors. The map f is infinitely renormalizable if there is an infinite sequence I_n of such intervals. It is very easy to see that the intervals I_n are nested and, by the non-existence of wandering intervals, it follows that the intersection of the orbits of I_n is a Cantor set that coincides with the closure of the critical orbit. The first major result in renormalization theory, that the *a priori* bounds are real, was obtained by Sullivan using the C^2 Koebe distortion lemma: if k_n is the period of the interval I_n then the map f^{k_n-1} has uniformly bounded distortion on an interval $J_n \supset f(I_n)$ that is mapped diffeomorphically over I_n. The crucial point here is that this bound on the distortion is independent of n and in fact becomes independent even of f for sufficiently large n. This implies in particular that the closure of the critical orbit has zero Lebesgue measure.

As we have mentioned before, the study of the distortion of the cross-ratio was introduced to obtain dynamical properties of smooth one-dimensional maps that do not have a negative Schwarzian derivative. With the intense development of these ideas, we have come back to the nice situation of maps with a negative Schwarzian derivative, if we restrict our attention to small scales. In fact, if c is a critical point of f that is not in the basin of a periodic attractor then there is a neighborhood U of c such that, for any point $x \in X$ and any integer $n \geq 0$ for which $f^n(x) \in U$, the Schwarzian derivative of f^{n+1} is negative at x. This result was obtained in the unimodal case in [Ko2] and extended to the multimodal situation in [SV]. It was used in the unimodal case in [GSS] to prove the following result: if f is a C^3 unimodal map with a non-flat critical point then there exists a neighborhood U of the critical point such that the first return map of f to U is real-analytically conjugate to a map with negative Schwarzian derivative.

As we have mentioned before, the Schwarzian derivative estimates were not used in Denjoy's theorem for circle diffeomorphisms. Neither were they mentioned in the celebrated theorem of M. Herman [He] on the smoothness of the conjugacy with rotation for circle diffeomorphisms whose rotation numbers satisfy a Diophantine condition. However, in the same paper Herman uses some careful arguments to control the distortion, which are related to the chain rule for the Schwarzian derivative. Very recently, the distortion of the cross-ratio was heavily in used in [KT] to give a new and very simple proof of an improved version of Herman's theorem.

6.2 Cross-ratios and *a priori* bounds

Having surveyed the developments, old and new, of cross-ratio distortion tools and their use in dynamics, we will present a simple application in a little more detail. Of course, much more can be found in [MS]. The application we have in mind is to establish *a priori* bounds for C^2 critical circle homeomorphisms. Our exposition here is taken from [dF2].

6.2.1 *Poincaré length and Koebe's principle*

Let us reintroduce the distortion tools to be used below, in a slightly different language. Following Sullivan, we define the *Poincaré density* of an open interval $I = (a, b) \subseteq \mathbb{R}$ to be

$$\rho_I(x) = \frac{b - a}{(x - a)(b - x)}$$

Integrating $\rho_I(x)\, dx$ we get the *Poincaré metric* on I. Thus, the Poincaré length of $J = (c, d) \subseteq I$ is

$$\ell_I(J) = \int_J \rho_I(x)\, dx = \log Cr[I, J],$$

where $Cr[I, J] = (a-d)(c-b)/(a-c)(d-b)$ is the cross-ratio of the four points a, b, c, d. If $f : I \to I^*$ is a diffeomorphism then the derivative of f measured with respect to the Poincaré metrics in I and I^*, namely

$$D_I f(x) \; = \; f'(x) \, \frac{\rho_{I^*}(f(x))}{\rho_I(x)} \; ,$$

is called the *Poincaré distortion* of f. It is identically equal to unity if f is Möbius, in which case f preserves cross-ratios. Now consider the symmetric function $\delta_f : I \times I \to \mathbb{R}$ given by

$$\delta_f(x, y) \; = \; \begin{cases} \log \dfrac{f(x) - f(y)}{x - y} \, , & x \neq y \, , \\[2mm] \log f'(x) \, , & x = y \, . \end{cases}$$

Then an easy calculation shows that

$$\log D_I f(x) \; = \; \delta_f(x, x) - \delta_f(a, x) - \delta_f(x, b) + \delta_f(a, b) \, . \tag{6.1}$$

Note that when f is C^3 its Poincaré distortion is controlled by the second-order mixed derivative of δ_f, since in that case

$$\log D_I f(t) \; = \; \int \int_Q \frac{\partial^2}{\partial x \, \partial y} \delta_f(x, y) \, dx dy \; ,$$

where Q is the square $(a, t) \times (t, b)$. Moreover, when $(x, y) \to (t, t)$ the integrand above becomes $-6Sf(x)$, where Sf is the Schwarzian derivative of f. This is consistent with the fact that maps with negative Schwarzian derivative increase the Poincaré metric and consequently decrease cross-ratios. Now, for C^2 mappings we have the following infinitesimal version, originally due to Sullivan, of a result of de Melo and van Strien [MS].

Lemma 6.2.1 Let $f : \overline{I} \to \mathbb{R}$ be a C^2 diffeomorphism onto its image. Then there exists a gauge function σ, depending only on the C^2 norm of f, such that $\nabla \delta_f$ is σ-Hölder, i.e.

$$|\nabla \delta_f(z_1) - \nabla \delta_f(z_2)| \; \leq \; \sigma(|z_1 - z_2|) \tag{6.2}$$

for all z_1 and z_2 in $\overline{I} \times \overline{I}$. In particular, $\log D_I f(x) \leq |x - a| \sigma(|x - a|)$ for all x in I.

Proof We prove (6.2), replacing the gradient by $\partial_x \delta_f$ and σ by some gauge function σ_x. We have

$$\partial_x \delta_f(x, y) \; = \; \frac{f'(x)(x - y) - f(x) + f(y)}{(f(x) - f(y))(x - y)} \; . \tag{6.3}$$

Now let $\mu : \{(x, h) \in I \times \mathbb{R} : a \le x + h \le b\} \to \mathbb{R}$ be given by

$$\mu(x, h) = \begin{cases} \dfrac{f(x + h) - f(x)}{h}, & h \ne 0, \\[2ex] f'(x), & h = 0. \end{cases}$$

Then μ is C^1 provided that f is C^2, and so we can write (6.3) as

$$\partial_x \delta_f(x, y) = \frac{\mu(x, 0)(x - y) - \mu(x, y - x)(x - y)}{\mu(x, y - x)(x - y)^2} \tag{6.4}$$

$$= \frac{\partial_h \mu(x, \vartheta)}{\mu(x, y - x)}, \tag{6.5}$$

for some $0 \le \vartheta \le |x - y|$. Let $m = \inf |\mu(x, h)| > 0$ and $M = \sup |\partial_h \mu(x, h)|$, both depending only on the C^1 norm of μ. Then if $z_i = (x_i, y_i) \in I \times I$ we have from (6.4)

$$|\partial_x \delta_f(z_1) - \partial_x \delta_f(z_2)| \le \frac{M}{m^2} |\mu(x_1, y_1 - x_1) - \mu(x_2, y_2 - x_2)| .$$

Thus, writing $\varphi(z) = \mu(x, y - x)$, we can take

$$\sigma_x(t) = \sup_{|z_1 - z_2| \le t} \frac{M}{m^2} |\varphi(z_1) - \varphi(z_2)| .$$

We define σ_y for $\partial_y \delta_f$ in the same way. Then the sum $\sigma = \sigma_x + \sigma_y$ satisfies (6.2). Finally, from (6.1) and the mean-value theorem, we have

$$\log D_I f(x) \le |x - a| \sigma_x(|x - a|) \le |x - a| \sigma(|x - a|) ,$$

and the lemma is proved.

$\square$

We note that the above definitions and lemma 6.2.1 still make sense when I is an interval in any Riemannian 1-manifold. Therefore we have the following result. Recall that a circle homeomorphism without periodic points is called *minimal* if the ω-limit set of any point is the whole circle.

Lemma 6.2.2 *Let f be a minimal C^2 circle homeomorphism, let N be a positive integer and let $I = I(N) \subseteq S^1$ be an interval such that: (a) $I, f(I), f^2(I), \ldots, f^N(I)$ are k-quasidisjoint for some $k > 0$ independent of N; (b) f restricted to each $f^i(I)$ is a diffeomorphism with C^2 norm uniformly bounded from below. Then the Poincaré distortion of f^N on I is bounded independently of N and goes to zero as $N \to \infty$.*

Proof The Poincaré distortion satisfies a chain rule. Therefore, if x is in I, lemma 6.2.1 yields

$$|\log D_I f^N(x)| = \left| \sum_{i=0}^{N-1} \log D_{f^i(I)} f(f^i(x)) \right|$$
$$\leq \sum_{i=0}^{N-1} \sigma(|f^i(I)|)\,|f^i(I)|$$
$$\leq \sigma(\ell_N) \sum_{i=0}^{N-1} |f^i(I)|\,,$$

where $\ell_N = \max_{0 \leq i < N} |f^i(I)|$. This last sum is bounded by k, while ℓ_N is also bounded independently of N. Since f is minimal there are no wandering intervals and therefore ℓ_N goes to zero as N goes to ∞. $\square$

This lemma tells us that, at small scales, long compositions of uniformly C^2 diffeomorphisms defined over quasidisjoint intervals are nearly projective. Now the *Koebe principle* (lemma 6.1.1) says that if a diffeomorphism is nearly projective over an interval I then, in a small subinterval J with definite space inside I, that diffeomorphism is in fact almost linear. The *space* $s(I, J)$ of J inside I is by definition the ratio of the length of the smaller of the two components of $I \setminus J$ and the length of J. We note that a C^2 version of Koebe's principle can be stated as follows.

Lemma 6.2.3 *Let* $f : I \to \mathbb{R}$ *be a* C^2 *diffeomorphism onto its image, and let* $J \subseteq I$ *be such that* $s(I, J) > 0$. *Then there exists a constant* C *depending only on* $s(I, J)$ *and the Poincaré distortion of* f *such that*

$$\left| \log \frac{f'(x)}{f'(y)} \right| \leq C|x - y|$$

for all x *and* y *in* J. *Moreover, for a fixed space,* C *goes to zero with the Poincaré distortion.*

Proof See [MS, Chapter IV]. $\square$

6.2.2 A priori bounds for circle maps

Now we use the above ideas to give a brief sketch of the proof of *a priori* bounds for C^2 critical circle maps. Let $f : S^1 \to S^1$ be an orientation-preserving C^2 homeomorphism, and let α be its rotation number. Only irrational α values will matter to us.

Let us agree to call f a *critical circle map* if f is a local C^2 diffeomorphism at all points except one, the critical point, around which f is locally C^2 conjugate to a *power law map* of the form $x \mapsto x|x|^{s-1} + a$. Here $s > 1$ is called the *type* of the critical point (the type $s = 3$, or *cubic* type, is the generic case). Let us denote the critical point by $c \in S^1$. Also, let $\{q_n\}_{n \geq 0}$ be the sequence of return times of the forward orbit of c to itself. This sequence turns out to be the same as the sequence of denominators of the rational numbers obtained by successive truncations of the continued fraction expansion of α. For each $n \geq 0$ let J_n be the closed interval on the circle with endpoints $f^{q_n}(c)$ and $f^{q_{n+1}}(c)$ that contains c. Then c divides J_n into two intervals, I_n with endpoint $f^{q_n}(c)$ and I_{n+1} with endpoint $f^{q_{n+1}}(c)$. The length ratio $s_n(f) = |I_{n+1}|/|I_n|$ is called the *nth scaling ratio* of f.

Theorem 6.2.4 *If $0 < \alpha < 1$ is an irrational number then there exist constants C_1 and C_2 such that:*

(i) *if f is a critical circle mapping having rotation number α then for all sufficiently large n we have $|I_{n+1}| \geq C_1 |I_n|$;*

(ii) *if f and g are critical circle mappings having rotation number α then for all sufficiently large n we have $|s_n(f)/s_n(g) - 1| \leq C_2$.*

We will only sketch the proof of this theorem. There is no loss of generality in assuming from the beginning that the critical circle map f is equal to a power law map in a small neighborhood of the critical point c. The first return map to J_n consists of f^{q_n} restricted to I_{n+1} and $f^{q_{n+1}}$ restricted to I_n. This pair is called the *nth renormalization* of f. It turns out that proving theorem 6.2.4 is tantamount to bounding the C^1 norms of these renormalizations (after we rescale both maps by a linear map taking I_n onto an interval of unit size) by a constant depending on f but not on n, which becomes independent even of f for large n. The key point is to get uniform space around the two intervals containing the critical value of f, namely $f(I_n)$ and $f(I_{n+1})$. Once this is accomplished, lemmas 6.2.2 and 6.2.3 give C^1 control of the renormalizations of f independently of n. For this purpose consider the collections

$$\mathcal{A}_n = \{I_n, f(I_n), f^2(I_n), \ldots, f^{q_{n+1}-1}(I_n)\}$$

and

$$\mathcal{B}_n = \{I_{n+1}, f(I_{n+1}), f^2(I_{n+1}), \ldots, f^{q_n-1}(I_{n+1})\} \ .$$

We have the following combinatorial facts, whose proofs are left as exercises.

Lemma 6.2.5 *For each $n \geq 0$, the union $\mathcal{P}_n = \mathcal{A}_n \cup \mathcal{B}_n$ is a partition of S^1.* □

This partition is sometimes referred to as the nth dynamical partition of f.

Lemma 6.2.6 *For each i in the range $1 \leq i \leq q_{n+1} - 1$, the inverse composition $f^{-i+1} : f^i(I_n) \to f(I_n)$ extends as a diffeomorphism to an interval $J_{i,n}$ containing $f^i(I_n)$ and its two nearest neighbors in $\mathcal{P}_n$.* □

The fundamental observation of Swiatek in [Sw] is that the smallest interval in $\mathcal{P}_n$ already has universal space around itself. More precisely, we may assume without loss of generality that $f^i(I_n) \in \mathcal{A}_n$ is the smallest interval in $\mathcal{P}_n = \mathcal{A}_n \cup \mathcal{B}_n$. Then by the definition of the space s of a subinterval we have

$$s(J_{i,n}, f^i(I_n)) \geq 1 \ ,$$

where $J_{i,n}$ is given by lemma 6.2.6. Using lemma 6.2.2, we can transfer this space to space around $f(I_n)$ as follows. We view the composition f^{-i+1} as being made up of factors of two types. There are *bounded factors*, namely those whose domains are far from the critical point and which have uniformly bounded C^2 norms, and there are *singular factors*, namely those whose domains fall inside a fixed neighborhood of the critical point. Since the singular factors are local inverses to a power law map, they have positive Schwarzian derivative and therefore can only *increase* space. The subcompositions of two singular factors are made up of bounded factors and therefore distort space by an additive uniformly bounded amount, according to lemma 6.2.2. Hence the whole composition has a uniformly bounded distortion of space, which gives us space for $f(I_n)$ inside $f^{-i+1}(J_{i,n})$. This fact plus a similar argument produces space around $f(I_{n+1})$ also. Finally, we can use lemma 6.2.3 to obtain that the C^1 norm of $f^{q_{n+1}}$ restricted to I_n is uniformly bounded, thereby proving (i) and (ii) of the theorem. Lemma 6.2.3 does not apply immediately to the whole composition, for not all factors in it are bounded. It is necessary to segregate the singular factors from the bounded factors, shuffling them apart by conjugating the bounded ones by the singular ones. This can be safely done because such conjugations are bounded operators in the space of C^2 diffeomorphisms with the C^1 topology.

Remark 6.2.7 *More generally, one can show that, for critical circle maps of class C^r with $r \geq 3$, successive renormalizations are bounded in the C^{r-1} topology. This requires a technical approximation lemma involving long compositions. For these results and much more, see [dFM1].*

Exercises

6.1   Prove the chain rules for the non-linearity and the Schwarzian derivative.

6.2   Verify that the Poincaré metric in $(-1, 1)$, as given in the text, indeed agrees with the restriction to $(-1, 1)$ of the hyperbolic metric of the unit disk.

6.3   Let $T \in PSL(2, \mathbb{R})$ be orientation preserving on the real line. Show that

$$\delta_T(x, y) = \tfrac{1}{2} \left(\log T'(x) + \log T'(y) \right) .$$

Deduce that, for every interval I in the domain of T, the Poincaré distortion of $T : I \to T(I)$ is identically zero.

6.4   Prove the following generalized version of Koebe's non-linearity principle (see [dFM1, p. 376]). Given $B, \tau > 0$, there exists $K_{\tau,B} > 0$ such that, if $f : [-\tau, 1+\tau] \to \mathbb{R}$ is a C^3 diffeomorphism into the reals and $Sf(t) \geq -B$ for all $t \in [-\tau, 1 + \tau]$, then we have

$$\left| \frac{f''(x)}{f'(x)} \right| \leq K_{\tau,B}$$

for all $0 \leq x \leq 1$. Show also that $K_{\tau,B} \to 2/\tau$ as $B \to 0$ (this recovers the classical Koebe non-linearity principle). (*Hint:* Write $y = Nf$ and apply an ODE comparison theorem to the solutions of the differential inequality $y' - \tfrac{1}{2}y^2 \geq -B$.)

6.5   Let $f : I \to f(I) \subseteq \mathbb{R}$ be a C^3 diffeomorphism without fixed points (I being a closed interval on the line). Show that if $Sf(x) < 0$ for all $x \in I$ then there exists a unique $x_0 \in I$ such that $|f(x_0) - x_0| \leq |f(x) - x|$ for all $x \in I$.

6.6   Let $\mathcal{P}_n$ be the nth dynamical partition of a critical circle map. For each $n \geq 1$, let

$$S_n \;=\; \sum_{I \in \mathcal{P}_n \setminus \{I_n, I_{n+1}\}} \left(\frac{|I|}{d(c, I)} \right)^2 ,$$

where $d(c, I)$ denotes the distance between the critical point and the interval I. Using the *a priori* bounds, prove that the sequence $\{S_n\}$ is bounded by a constant depending only on f. (This result is from [dFM1].)

Appendix
Riemann surfaces and Teichmüller spaces

A.1 Riemann surfaces

A *Riemann surface* is a connected, complex, one-dimensional manifold. In more technical parlance, a Riemann surface is a Hausdorff, second-countable, connected topological space X on which there is a *holomorphic atlas* $\{(U_\alpha, \varphi_\alpha)\}$, i.e. a family of charts $\varphi_\alpha : U_\alpha \to \mathbb{C}$, with $\{U_\alpha\}$ an open covering of X, which are homeomorphisms onto their images and whose *chart transitions*

$$h_{\alpha\beta} = \varphi_\beta \circ \varphi_\alpha^{-1} : \varphi_\alpha(U_\alpha \cap U_\beta) \to \varphi_\beta(U_\alpha \cap U_\beta)$$

are *bi-holomorphic maps*. Every Riemann surface is *a fortiori* an orientable real two-dimensional surface, because we have $\mathrm{Jac}(h_{\alpha\beta}) = |h'_{\alpha\beta}|^2 > 0$ for all indices α, β, so that a holomorphic atlas is necessarily oriented. Two holomorphic atlases on the same topological space X are said to be compatible if their union is also a holomorphic atlas on X. This is clearly an equivalence relation; the equivalence classes are the *Riemann surface structures*, or *holomorphic structures*, on X. Alternatively, one can think of a holomorphic structure on X as a *maximal* holomorphic atlas on X, where by "maximal" we mean that such an atlas is not properly contained in any other atlas compatible with it.

Natural examples of Riemann surfaces abound. In fact, Riemann surfaces are some of the most ubiquitous objects in all mathematics. Some basic examples will be given below.

A continuous map $f : X \to Y$ between two Riemann surfaces X, Y is said to be *holomorphic* if for all charts (U, φ) and (V, ψ) on X and Y respectively, with $f(U) \subseteq V$, we have that $\psi \circ \varphi^{-1} : \varphi(U) \to \psi(V)$ is holomorphic in the usual sense (as a map between open sets in the complex plane). One defines a *quasiconformal* map from X to Y in pretty much the same way, simply replacing the word holomorphic by the word quasiconformal in the definition just given.

If a holomorphic map $f : X \to Y$ happens to be invertible, its inverse $f^{-1} : Y \to X$ is holomorphic as well (this follows from the inverse function theorem for complex analytic functions). In this case the map f is said to be *bi-holomorphic*, or a *conformal equivalence*.

A bi-holomorphic map $f : X \to X$ of a Riemann surface X onto itself is called an *automorphism* of X. The automorphisms of X form a group under composition, denoted by $\mathrm{Aut}(X)$.

A.1.1 Examples

Here are some of the main examples of Riemann surfaces. The unproved assertions made below are left as exercises.

(1) *Simply connected Riemann surfaces*: the complex plane $\mathbb{C}$, the Riemann sphere $\widehat{\mathbb{C}} = \mathbb{C} \cup \{\infty\}$ (using stereographic projection to build charts), the unit disk $\mathbb{D}$ and the upper half-plane $\mathbb{H}$. These last two are conformally equivalent, via a Möbius transformation. We have also the following facts:

    (a) $\mathrm{Aut}(\mathbb{C}) = \{z \mapsto az + b \,|\, a \in \mathbb{C}^*, b \in \mathbb{C}\}$ (the affine group);

    (b) $\mathrm{Aut}(\widehat{\mathbb{C}}) \simeq PSL(2,\mathbb{C})$ (the Möbius group, see Chapter 3);

    (c) $\mathrm{Aut}(\mathbb{D}) \simeq \mathrm{Aut}(\mathbb{H}) \simeq PSL(2,\mathbb{R})$.

(2) *Any open subset of the above examples*: in other words, any open subset of the Riemann sphere is a Riemann surface. Important special cases include the annuli $A_{r,R} = \{z : r < |z| < R\}$, the punctured plane $\mathbb{C}^*$, the doubly punctured plane $\mathbb{C}^{**}$ etc.

(3) *Complex tori* (figure A.1): these arise as quotients $\mathbb{T}_\lambda = \mathbb{C}/(\mathbb{Z} \oplus \lambda\mathbb{Z})$ of the additive group $\mathbb{C}$ by a free abelian subgroup on two generators $\{1, \lambda\}$, where $\mathrm{Im}\,\lambda > 0$. The reader can build charts by hand or wait for the next example below. (*Exercise* what is $\mathrm{Aut}(\mathbb{T}_\lambda)$?)

(4) *Quotient Riemann surfaces*: if $\widehat{X}$ is a Riemann surface and $\Gamma \subseteq \mathrm{Aut}(\widehat{X})$ is a torsion-free discontinuous subgroup of automorphisms then the quotient space $X = \widehat{X}/\Gamma$ is a Riemann surface. Included here are the complex tori of our previous example, as well as the so-called *hyperbolic* Riemann surfaces, i.e. quotients of the form $X = \mathbb{H}/\Gamma$ where $\Gamma \subseteq PSL(2,\mathbb{R})$ is discontinuous and torsion-free. All compact Riemann surfaces of genus $g > 1$ arise

in this way; if X is such a surface, one can show that $\mathrm{Aut}(X)$ is a *finite* group.

From the point of view of *abstract* Riemann surface theory, these are all the examples that we care about. But in other areas of mathematics, one is often interested in *concrete realizations* of Riemann surfaces. Thus, in algebraic geometry, a Riemann surface arises essentially as the locus of points in $\mathbb{C} \times \mathbb{C}$ where a given complex polynomial in two variables vanishes. Such a viewpoint is closer in spirit to the one originally taken up by Riemann. This is just the tip of a huge iceberg; see for instance [GH, chapter 2].

A.1.2 The uniformization theorem

The famous *uniformization theorem* of Klein, Poincaré and Koebe – a version of which we proved in Chapter 3 – tells us that the above list of examples exhausts *all* Riemann surfaces up to conformal equivalence.

Theorem A.1.1 (Uniformization) *Every Riemann surface is, up to conformal equivalence, a quotient of the form $X = \widehat{X}/\Gamma$, where $\widehat{X}$ is either the complex plane $\mathbb{C}$, the Riemann sphere $\widehat{\mathbb{C}}$ or the upper half-plane $\mathbb{H}$, and where Γ is a torsion-free discontinuous group of conformal self-maps of $\widehat{X}$.*

A complete proof of this theorem can be found in [FK, chapter IV]. A full treatment should include the so-called Koebe–Maskit theorem, which states that if X is a Riemann surface topologically equivalent to a plane domain then X is *conformally* equivalent to it. Note that the uniformization theorem says in particular that any Riemann surface has either the upper half-plane, the complex plane or the Riemann sphere as its (holomorphic) universal cover.

We remark that the list of examples given in subsection A.1.1 contains many redundancies. For instance:

(i) any two simply connected, proper, open subsets of the complex plane are conformally equivalent, by the Riemann mapping theorem;

(ii) if $\lambda, \mu \in \mathbb{H}$ lie in the same orbit, by the modular group $PSL(2, \mathbb{Z})$, then the corresponding tori $\mathbb{T}_\lambda$ and $\mathbb{T}_\mu$ are conformally equivalent;

(iii) if Γ_1 and Γ_2 are (discontinuous torsion-free) *conjugate* subgroups of $PSL(2, \mathbb{R})$, in other words, if there exists $\gamma \in PSL(2, \mathbb{R})$ such that $\Gamma_2 = \gamma\Gamma_1\gamma^{-1}$, then the corresponding Riemann surfaces $X_1 = \mathbb{H}/\Gamma_1$ and $X_2 = \mathbb{H}/\Gamma_2$ are conformally equivalent.

A Riemann surface X is said to be *conformally finite* if X is conformally equivalent to a compact Riemann surface punctured at a finite set of points. More precisely, X has finite conformal type (g, n) if there exist a compact Riemann surface $\widehat{X}$ of genus g and a finite set $F \subset \widehat{X}$ with exactly n points, together with a conformal embedding $\iota : X \to \widehat{X}$ such that $\iota(X) = \widehat{X} \setminus F$. We leave to the reader the task of verifying that the conformal type (g, n) is well defined. One refers to F as the set of *punctures*. The vast majority of such conformally finite surfaces are hyperbolic surfaces. The exceptions are the cases $g = 0$, $n \leq 2$, $g = 1$, $n = 0$. Finally, the *ideal boundary* ∂X of a Riemann surface X can be defined as follows. If X is conformally finite, let ∂X be the set of punctures of X. If X is not conformally finite, it is hyperbolic and we can write $X = \mathbb{H}/\Gamma$, with Γ a Fuchsian group as in the uniformization theorem. Let $\Lambda \subset \widehat{\mathbb{R}} = \mathbb{R} \cup \infty$ be the *limit set* of the Fuchsian group Γ. This is a closed set with empty interior and it is Γ-invariant, hence so is its complement. Let $\partial X = (\widehat{\mathbb{R}} \setminus \Lambda)/\Gamma$.

A.1.3 Beltrami and quadratic differentials

As in the case of general differentiable manifolds, one can consider various tensor bundles over a given Riemann surface X, whose sections are relevant objects from a geometric (or holomorphic) viewpoint. The basic building block is the holomorphic cotangent bundle of X, also called the *canonical line bundle* of X and denoted by $\mathcal{K}_X$. The two most important objects for our purposes are the following.

1 Beltrami differentials

A Beltrami differential on a Riemann surface X is a tensor of type $(-1, 1)$ over X, i.e. a section of the bundle $\mathcal{K}_X^{-1} \otimes \overline{\mathcal{K}_X}$. To be more concrete, let $\{(U_\alpha, \varphi_\alpha)\}$ be a maximal atlas over X. Then a Beltrami differential μ on X is given by a collection of local functions μ_α defined on $\varphi_\alpha(U_\alpha)$ satisfying the transition relations

$$\mu_\alpha(z) = \mu_\beta(h_{\alpha\beta}(z)) \frac{\overline{h'_{\alpha\beta}(z)}}{h'_{\alpha\beta}(z)} \tag{A.1}$$

for all $z \in \varphi_\alpha(U_\alpha \cap U_\beta)$, where $h_{\alpha\beta} = \varphi_\beta \circ \varphi_\alpha^{-1}$ are the chart transitions. Note in particular that $|\mu|$ defines a function on X. We require the essential supremum $\|\mu\|_\infty$ of this function to be finite. Hence the space of all Beltrami differentials on X can be identified with the Banach space $B(X) = L^\infty(X, \mathbb{C})$. We denote by $M(X)$ the unit ball in $B(X)$, i.e. $M(X) = \{\mu \in B(X) : \|\mu\|_\infty < 1\}$.

Every $\mu \in M(X)$ gives rise to a (new) conformal structure on X. This can be seen as follows. Let μ_α represent μ on the chart $(U_\alpha, \varphi_\alpha)$. By the measurable Riemann mapping theorem (see Chapter 4), there exists a quasiconformal map

$$f_\alpha : \varphi_\alpha(U_\alpha) \to f_\alpha \circ \varphi_\alpha(U_\alpha) \subseteq \mathbb{C}$$

such that $\bar{\partial} f_\alpha(z) = \mu_\alpha(z) \, \partial f_\alpha(z)$ for all $z \in \varphi_\alpha(U_\alpha)$. Let $\mathcal{A}^\mu$ be the atlas on X whose charts are $(U_\alpha, f_\alpha \circ \varphi_\alpha)$. Using the transition relations (A.1) and the composition formula for complex dilatations, it is not difficult to show (exercise) that the new overlaps

$$\tilde{h}_{\alpha\beta} = f_\beta \circ \varphi_\beta \circ (f_\alpha \circ \varphi_\alpha)^{-1} = f_\beta \circ h_{\alpha\beta} \circ f_\alpha^{-1}$$

are *holomorphic*. Hence $\mathcal{A}^\mu$ determines a new conformal structure on X. The resulting Riemann surface is denoted by X^μ. The reader can check, as an exercise, that the identity map id : $X \to X$, viewed as a map $X \to X^\mu$, is quasiconformal.

2 Quadratic differentials

A quadratic differential on a Riemann surface X is a tensor of type $(2, 0)$ on X, i.e. an expression of the form $\phi(z) \, dz^2$ in local coordinates. More precisely, a quadratic differential on X is a section of the bundle $\mathcal{K}_X^2 = \mathcal{K}_X \otimes \mathcal{K}_X$. Again, in terms of local charts $(U_\alpha, \varphi_\alpha)$ on X, a quadratic differential ϕ on X is given by a collection of functions $\phi_\alpha : \varphi_\alpha(U_\alpha) \to \mathbb{C}$ such that

$$\phi_\alpha(z) = \phi_\beta(h_{\alpha\beta}(z)) \left(h'_{\alpha\beta}(z)\right)^2 \tag{A.2}$$

for all $z \in \varphi_\alpha(U_\alpha \cap U_\beta)$, where $h_{\alpha\beta}$ are the chart transitions as before.

A quadratic differential gives rise to various structures on X. Thus, from (A.2) we see that $|\phi(z)| \, dxdy$ is a *measure* on X. A quadratic differential $\phi \, dz^2$ on X is said to be *integrable* if its measure is finite, in other words if

$$\|\phi\| = \iint_X |\phi| \, dxdy$$

is finite. The space of integrable quadratic differentials is a Banach space, the dual to the space of Beltrami differentials on X. (This duality is crucial in the infinitesimal theory of Teichmüller spaces, but it will not concern us here.)

We are interested here only in the subspace consisting of *holomorphic quadratic differentials* on X, denoted $Q(X)$. The fundamental result about $Q(X)$ that will be needed later is the following.

Theorem A.1.2 *Let X be a conformally finite Riemann surface of the type (g,n). Then $Q(X)$ is a finite-dimensional complex vector space, and in fact*

$$
\dim_{\mathbb{C}} Q(X) = \begin{cases} 0 & \text{if } g = 0 \text{ and } n \leq 2, \\ 1 & \text{if } g = 1 \text{ and } n = 0, \\ 3g - 3 + n & \text{in all other cases}. \end{cases}
$$

This theorem is a corollary of a central result in Riemann surface theory, the *Riemann–Roch theorem*. See [FK, chapter III] for a complete proof. Note in particular that for compact Riemann surfaces of genus $g \geq 2$ the theorem asserts that $\dim_{\mathbb{C}} Q(X) = 3g - 3$. See figure 7.1 for an illustration of a compact Riemann surface.

A.1.4 Moduli spaces

How does one go about describing all Riemann surfaces of a given topological type, up to conformal equivalence? By asking this question, we are led quite naturally to the notion of Riemann's *moduli space*. The Riemann moduli space of a given Riemann surface X is the space of all conformal equivalence classes of Riemann surfaces that are *quasiconformally* equivalent to X.

To put it more formally, let us agree that a *marked Riemann surface of type X* is a pair (f, Y) where Y is a Riemann surface and $f : X \to Y$ is a quasiconformal homeomorphism. Two marked Riemann surfaces of type X, say (f_1, Y_1) and (f_2, Y_2), are said to be equivalent, $(f_1, Y_1) \approx (f_2, Y_2)$, if there exists a conformal bi-holomorphic map $c : Y_1 \to Y_2$ such that $c \circ f_1 = f_2$. This is clearly an equivalence relation. The Riemann moduli space of X, denoted $\mathcal{M}(X)$, is the quotient space of the space of marked Riemann surfaces of type X and this equivalence relation: $\mathcal{M}(X) = \{(f, Y)\}/\approx$.

A.2 Teichmüller theory

The notion of Teichmüller space originated before the second world war in the works of O. Teichmüller, but the theory of Teichmüller spaces as we know it today really flourished in the hands of L. Ahlfors and L. Bers and their school (see [Be] for a useful survey). In such a short appendix we can hardly do justice to this beautiful and powerful theory. We limit ourselves to the statements, and sometimes proofs, of the essential results that were needed in the proof of the Bers–Royden theorem.

A.2.1 Teichmüller spaces

Let X be a Riemann surface. We are especially interested in the case when X is hyperbolic: namely $X = \mathbb{H}/\Gamma$, the quotient of the upper half-plane and a discrete subgroup $\Gamma \subset PSL(2, \mathbb{R})$ of the group of isometries in the Poincaré metric.

Let us consider the class of all *pairs* of the form (f, Y) where Y is a Riemann surface and $f : X \to Y$ is a qc-homeomorphism. We say that two such pairs (f_1, Y_1) and (f_2, Y_2) are *equivalent*, $(f_1, Y_1) \sim (f_2, Y_2)$, if there exists a *conformal* map $c : Y_1 \to Y_2$ such that

$$f_2^{-1} \circ c \circ f_1 : X \to X$$

is homotopic to id_X relative to the *ideal boundary* of X. It is easily seen that this is indeed an equivalence relation. We also leave as an exercise for the reader to check that, when X is hyperbolic, $(f_1, Y_1) \sim (f_2, Y_2)$ if and only if there exists a quasiconformal homeomorphism $\psi : \mathbb{H} \to \mathbb{H}$ such that $\psi|_{\partial \mathbb{H}} = \mathrm{id}_{\partial \mathbb{H}}$ and such that the diagram

$$
\begin{CD}
\mathbb{H} @>{\psi}>> \mathbb{H} \\
@V{\pi}VV @VV{\pi}V \\
X @>>{f_2^{-1} \circ c \circ f_1}> X
\end{CD}
$$

commutes (here $\pi : \mathbb{H} \to X$ denotes the canonical quotient map).

Definition A.2.1 *The set of all equivalence classes of pairs (f, Y) under the above equivalence relation is the* Teichmüller space *of the Riemann surface X, and is denoted* Teich(X).

The equivalence class of a pair (f, Y) will be denoted by $[f]$ (we shall ignore Y in representing such elements of Teich(X) because $Y = f(X)$ is determined by f).

There is a natural way of making Teich(X) into a metric space. If $[f_1]$ and $[f_2]$ are any two elements of Teich(X), we let their *Teichmüller distance* be given by

$$d_T([f_1], [f_2]) = \inf_f \log K_f ,$$

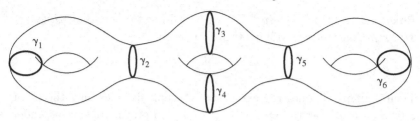

Fig. A.1. A compact Riemann surface of type (3) and its parts decomposition.

where f ranges over all qc-homeomorphisms $Y_1 \to Y_2$ that are equivalent to $f_2 \circ f_1^{-1}$.

A.2.2 The Bers embedding

Suppose that X is a hyperbolic Riemann surface, say $X = \mathbb{H}/\Gamma$, with Γ a discrete, torsion-free subgroup of $PSL(2, \mathbb{R})$ as before. Our goal in this section is to show that $\mathrm{Teich}(X)$ carries a natural complex structure.

This is elegantly achieved by the so-called *Bers embedding*. Let $\widetilde{B}(\Gamma)$ denote the space of holomorphic Γ-invariant quadratic differentials defined in the lower half-plane $\mathbb{H}^*$. An element of $\widetilde{B}(\Gamma)$ is identified with a holomorphic function $\varphi : \mathbb{H}^* \to \mathbb{C}$ such that

$$\varphi(\gamma z)\big(\gamma'(z)\big)^2 = \varphi(z)$$

for all $z \in \mathbb{H}^*$ and all $\gamma \in \Gamma$. We define $B(\Gamma) \subseteq \widetilde{B}(\Gamma)$ to be the subset of those φ such that

$$\|\varphi\| = \sup_{z \in \mathbb{H}^*} y^2 |\varphi(z)| < \infty \quad (z = x + iy) .$$

This defines a norm on $B(\Gamma)$, called the *Nehari norm*, and makes it into a normed complex vector space. The Nehari norm turns out to be complete (see exercise A.6), so $B(\Gamma)$ is in fact a complex Banach space.

Following Bers, we will show that $\mathrm{Teich}(X)$ embeds into $B(\Gamma)$ when $X = \mathbb{H}/\Gamma$. We need the following auxiliary results.

Lemma A.2.2 (Nehari–Kraus) *Let* $\phi : \mathbb{H}^* \to \widehat{\mathbb{C}}$ *be a univalent map. Then the holomorphic quadratic differential* $\varphi \, dz^2$ *given by*

$$\varphi = S\phi = \frac{\phi'''}{\phi'} - \frac{3}{2}\left(\frac{\phi''}{\phi'}\right)^2$$

has Nehari norm $\|\varphi\| < 3/2$.

Proof Let us fix $z = x + iy \in \mathbb{H}^*$; we would like to show that $|S\phi(z)| \leq 3/(2y^2)$. Consider the Möbius transformation

$$T(\zeta) = \frac{\zeta - \bar{z}}{\zeta - z}.$$

Then T maps $\mathbb{H}^*$ onto the exterior of the unit disk, sending the point z to ∞. Note that the inequality we want to prove is invariant under post-composition of ϕ by a Möbius transformation (because such post-composition does not change the Schwarzian). Hence we may suppose that $\phi(z) = \infty$ and that $\phi'(z) = 1/(T^{-1})'(\infty)$. Then the map $F = \phi \circ T^{-1}$ is univalent in the exterior of the unit disk and is tangent to the identity at ∞, say

$$F(w) = w + \frac{b_1}{w} + \frac{b_2}{w^2} + \cdots$$

By the area theorem (theorem 2.2.3) we have $\sum n|b_n|^2 \leq 1$, so $|b_1| \leq 1$ *a fortiori*. Now, a straightforward calculation yields

$$SF(w) = -\frac{6b_1}{w^4} + \cdots,$$

where the ellipses denote the terms in powers of $1/w$ higher than the fourth. Therefore, since $S\phi(\zeta) = SF(T(\zeta))(T'(\zeta))^2$ and $T'(\zeta) = -2y/(\zeta - z)^2$, we have

$$S\phi(\zeta) = (-6b_1 + \cdots)\frac{1}{T(\zeta)^4}\frac{4y^2}{(\zeta - z)^4}$$

$$= (-6b_1 + \cdots)\frac{4y^2}{(\zeta - \bar{z})^4}.$$

Taking the limit as $\zeta \to z$, we get

$$S\phi(z) = -\frac{6b_1}{4y^2},$$

and since $|b_1| \leq 1$, we are done.

$\square$

Lemma A.2.3 *Let $\phi : V \to \mathbb{C}$ be holomorphic on $V \subseteq \mathbb{C}$, and let ξ, η be two linearly independent solutions to $y'' + \frac{1}{2}\phi y = 0$. Then $f = \xi/\eta$ satisfies $Sf = \phi$.*

Proof By direct calculation, we have

$$f' = \left(\frac{\xi}{\eta}\right)' = \frac{\xi'\eta - \xi\eta'}{\eta^2}$$

as well as

$$f'' = \left(\frac{\xi}{\eta}\right)'' = \frac{-2\eta'(\xi'\eta - \xi\eta')}{\eta^3},$$

where we have used the fact that $\xi''\eta - \xi\eta'' = 0$. Hence the non-linearity of f is

$$\frac{f''}{f'} = -2\frac{\eta'}{\eta}.$$

Therefore

$$Sf = \left(\frac{f''}{f'}\right)' - \frac{1}{2}\left(\frac{f''}{f'}\right)^2 = -2\frac{\eta''}{\eta} = \phi.$$

$\square$

The following is by far the most delicate result that we will prove in this appendix, and it is known as the *Ahlfors–Weill section theorem*.

Theorem A.2.4 (Ahlfors–Weill) *Let φ be holomorphic in the lower half-plane $\mathbb{H}^*$, and suppose that its Nehari norm satisfies $\|\varphi\| < 1/2$. Consider the (harmonic) Beltrami coefficient μ given by $\mu(z) = -2y^2\varphi$ $(\bar{z})$ for $z \in \mathbb{H}$ and $\mu(z) \equiv 0$ everywhere else. Then the corresponding normalized solution to the Beltrami equation $f^\mu : \widehat{\mathbb{C}} \to \widehat{\mathbb{C}}$ satisfies $Sf^\mu = \varphi$ in $\mathbb{H}^*$. Moreover, if φ is Γ-invariant as a quadratic differential for some Fuchsian group Γ then so is μ as a Beltrami differential.*

Proof Let us first suppose that φ is holomorphic in a neighborhood of $\mathbb{H}^* \cup \{\infty\}$ in $\widehat{\mathbb{C}}$ and that $|z^4\varphi(z)|$ remains bounded as $|z| \to \infty$.

Let ξ and η be linearly independent solutions to $y'' = -\frac{1}{2}\varphi y$ as in lemma A.2.3, normalized in such a way that $\xi'\eta - \xi\eta' = 1$. Note that with this normalization ξ and η do not vanish simultaneously anywhere. Define $F : \widehat{\mathbb{C}} \to \widehat{\mathbb{C}}$ by

$$F(z) = \begin{cases} F^+(z) = \dfrac{\xi(\bar{z}) + (z - \bar{z})\xi'(\bar{z})}{\eta(\bar{z}) + (z - \bar{z})\eta'(\bar{z})} & \text{for } z \in \mathbb{H}, \\[2ex] F^-(z) = \dfrac{\xi(z)}{\eta(z)} & \text{for } z \in \overline{\mathbb{H}^*}. \end{cases}$$

Then F is holomorphic in the lower half-plane, and we know from lemma A.2.3 that $SF = \varphi$ there. Note that since φ is holomorphic across the real axis so are ξ and η, and one easily sees that $F^+(z) = F^-(z)$ for all $z \in \mathbb{R}$; from this it follows that F is continuous across the real axis.

Moreover, a direct computation yields

$$\frac{\overline{\partial} F(z)}{\partial F(z)} = -y^2 \varphi(\overline{z}) = \mu(z)$$

for all $z \in \mathbb{H}$. This shows that the Beltrami coefficient of F is equal to μ almost everywhere.

We now prove that F is indeed a homeomorphism. It suffices to show that F is a *local* homeomorphism at each point (because F is a map of the sphere into itself; see exercise A.7).

The Jacobian of F^+ is equal to $|\partial F^+(z)|^2(1 - |\mu(z)|^2)$, as another straightforward computation shows. Since $|\mu(z)| < 1$ for all $z \in \mathbb{H}$, to show that F^+ is a local homeomorphism at z it is enough to show that $\partial F^+(z) \neq 0$. By a direct calculation using the normalization condition on ξ and η, we get

$$\partial F^+(z) = \frac{1}{(\eta(\overline{z}) + (z - \overline{z})\eta'(\overline{z}))^2},$$

from which it follows that $\partial F^+(z)$ is never zero (at the potentially dangerous places where the denominator might vanish, one replaces F^+ by $1/F^+$). Hence F^+ is a local homeomorphism everywhere in a neighborhood of the upper half-plane in $\mathbb{C}$ (and it is orientation preserving there). Similarly, the Jacobian of F^- at every z in a neighborhood of $\mathbb{H}^*$ in $\mathbb{C}$ is equal to $|\partial F^-(z)|^2$. This time we have

$$\partial F^-(z) = \frac{1}{(\eta(z))^2}$$

and once again, by similar considerations to those used before, we deduce that this never vanishes and F^- is a local (orientation-preserving) homeomorphism. Since F^+ and F^- agree in the real axis, F is a local homeomorphism everywhere in the complex plane. We still need to check that F is a local homeomorphism in a neighborhood of infinity. Using the fact that $|\varphi(z)| = O(|z|^{-4})$ near infinity, it is not difficult to see (exercise) that ξ and η, being solutions to $y'' + \frac{1}{2}\varphi y = 0$, have the form

$$\xi(z) = az + b + O(|z|^{-1}),$$
$$\eta(z) = cz + d + O(|z|^{-1})$$

for z near infinity. Here a, b, c, d are complex constants with $ad - bc = 1$, so a and c cannot be both zero. From these facts it follows that $F^- = \xi/\eta$ is a local homeomorphism at ∞. Similarly, F^+ is a local homeomorphism at ∞ and therefore the same is true for F.

Thus we have proved that F is a quasiconformal homeomorphism and that its Beltrami coefficient is equal to μ. Hence there exists a Möbius transformation T such that $f^\mu = T \circ F$. But then f^μ is also holomorphic in the lower half-plane, and we have $Sf^\mu = SF = \varphi$ there. Note that if φ is Γ-invariant as a quadratic differential then μ is Γ-invariant as a Beltrami differential, and the last assertion in the statement of the theorem follows from exercise 4.9.

The final step in the proof is to remove the extra hypotheses we made on φ (namely, that it is holomorphic across the real axis and has a fourth-order zero at ∞). For each positive integer n, let T_n be the Möbius transformation

$$T_n(z) = \frac{nz - i}{iz + n} \, .$$

Then T_n maps the lower half-plane $\mathbb{H}^*$ onto a disk compactly contained in $\mathbb{H}^*$, and $T_n(z) \to z$ for all z as $n \to \infty$. Let $\varphi_n \, dz^2$ be the quadratic differential defined by

$$\varphi_n(z) = \varphi(T_n(z)) \left(T_n'(z) \right)^2 \, .$$

Note that φ_n is holomorphic in the open set $T_n^{-1}(\mathbb{H}^*)$, which is a neighborhood of $\mathbb{H}^*$ in the Riemann sphere. Moreover, φ_n has a fourth-order zero at ∞ (because T_n' has a second-order zero at that point). Hence each φ_n satisfies our extra hypotheses. Since T_n contracts the hyperbolic metric of $\mathbb{H}^*$, we have $|T_n'(z)| \le |y^{-1} \mathrm{Im}\, T_n(z)|$, and therefore

$$|y^2 \varphi_n(z)| \le |\mathrm{Im}\, T_n(z)|^2 |\varphi(T_n(z))| \, .$$

Taking the supremum over all $z \in \mathbb{H}^*$, we deduce that $\|\varphi_n\| \le \|\varphi\| < 1/2$ for all n. Hence each φ_n satisfies *all* the hypotheses of our theorem. Let μ_n be the Beltrami differential equal to $-2y^2 \varphi_n(\overline{z})$ for all $z \in \mathbb{H}$ and equal to zero everywhere else. Applying what has been proved so far, we know that for each n there exists a normalized quasiconformal homeomorphism $f_n : \widehat{\mathbb{C}} \to \widehat{\mathbb{C}}$ with Beltrami coefficient equal to μ_n and such that $Sf_n(z) = \varphi_n(z)$ for all $z \in \mathbb{H}^*$. Note that $\|\mu_n\|_\infty \le \|\mu(z)\|_\infty < 1$ for all n. In addition, since φ_n converges uniformly on compact subsets of the lower half-plane to φ, we see that $\mu_n(z) \to \mu(z)$ pointwise (almost everywhere) as $n \to \infty$, where $\mu(z) = -2y^2 \varphi(\overline{z})$ for $z \in \mathbb{H}$ and $\mu \equiv 0$ everywhere else. These facts show (see exercise A.8) that f_n converges (uniformly on compact subsets) to a quasiconformal homeomorphism $f : \widehat{\mathbb{C}} \to \widehat{\mathbb{C}}$ whose Beltrami coefficient is equal to μ. Moreover, $Sf_n = \varphi_n$ converges to Sf uniformly on compact subsets of the lower half-plane, so that $Sf = \varphi$, and this finishes the proof. □

Theorem A.2.5 (Bers) *There exists an embedding* β : Teich$(X) \to B(\Gamma)$ *whose image is a bounded open set in* $B(\Gamma)$.

Proof Let $\tau \in$ Teich(X), and let $\mu \in L^\infty(\mathbb{H})$ with $\|\mu\|_\infty < 1$ be a representative of τ (recall that μ is Γ-invariant). We want to associate with μ an element of $B(\Gamma)$. To do this, let $\widetilde{\mu} \in L^\infty(\widehat{\mathbb{C}})$ be the Beltrami coefficient given by $\widetilde{\mu} \equiv \mu$ on $\mathbb{H}$ and $\widetilde{\mu} \equiv 0$ everywhere else. Let f^μ : $\widehat{\mathbb{C}} \to \widehat{\mathbb{C}}$ be the (unique) normalized solution to the Beltrami equation

$$\overline{\partial} f^\mu \ = \ \widetilde{\mu} \partial f^\mu \ .$$

Note that $\widetilde{\mu}$ is Γ-invariant (because μ is) and therefore

$$f^\mu \circ \gamma = \gamma \circ f^\mu$$

for all $\gamma \in \Gamma$. Note also that $f^\mu|_{\mathbb{H}^*}$ is conformal, i.e. a univalent map. We define

$$\varphi^\mu \ = \ S(f^\mu|_{\mathbb{H}^*}) \ ,$$

the Schwarzian derivative of such a univalent map. Using the chain rule for the Schwarzian derivative, we see from

$$S(f^\mu|_{\mathbb{H}^*} \circ \gamma) \ = \ S(\gamma \circ f^\mu|_{\mathbb{H}^*})$$

that

$$S(f^\mu|_{\mathbb{H}^*}) \circ \gamma \, (\gamma')^2 \ = \ S(f^\mu|_{\mathbb{H}^*}) \ .$$

In other words, we have

$$\varphi^\mu \circ \gamma \, (\gamma')^2 \ = \ \varphi^\mu$$

for all $\gamma \in \Gamma$ and this shows that, indeed, $\varphi^\mu \in B(\Gamma)$.

Next, we need to show that φ^μ is independent of which representative μ of τ we choose. We claim that if $\nu \in L^\infty(\mathbb{H})$ with $\|\nu\|_\infty < 1$ is Γ-invariant and is equivalent to μ, then $f^\nu|_{\mathbb{H}^*} = f^\mu|_{\mathbb{H}^*}$. This clearly implies that $\varphi^\mu = \varphi^\nu$, which is what we want. To prove the claim, let $f_\mu : \mathbb{H} \to \mathbb{H}$ be the normalized solution to $\overline{\partial} f_\mu \ = \ \mu \, \partial f_\mu$ in the upper half-plane, and let f_ν be similarly defined. Then, since μ and ν are equivalent, we know that $f_\mu|_{\partial\mathbb{H}} = f_\nu|_{\partial\mathbb{H}}$. This allows us to define a normalized quasiconformal map $f : \widehat{\mathbb{C}} \to \widehat{\mathbb{C}}$ by

$$f \ = \ \begin{cases} (f_\nu)^{-1} \circ f_\mu & \text{on } \ \overline{\mathbb{H}}, \\ \text{id} & \text{on } \ \mathbb{H}^* \ . \end{cases}$$

Now let $g : \widehat{\mathbb{C}} \to \widehat{\mathbb{C}}$ be the map

$$g = f^\mu \circ f \circ (f_\mu)^{-1} .$$

This is a quasiconformal map (as the welding of two such maps). Note that g is actually *conformal* when restricted to $f^\mu(\mathbb{H}^*)$, mapping it onto $f^\nu(\mathbb{H}^*)$. Indeed,

$$g|_{f^\mu(\mathbb{H}^*)} = f^\nu \circ (f^\mu)^{-1} .$$

It is also conformal when restricted to $f^\mu(\mathbb{H})$, since

$$g|_{f^\mu(\mathbb{H})} = (f^\nu \circ f_\nu^{-1}) \circ (f_\mu \circ (f^\mu)^{-1})$$

and both $f^\nu \circ f_\nu^{-1}$ and $f_\mu \circ (f^\mu)^{-1}$ are conformal, because $f^\mu|_{\mathbb{H}}$ and f_μ have the same Beltrami coefficient, and similarly for f^ν and f_ν. Hence g is conformal almost everywhere, and therefore conformal. Since it is also normalized it must be the identity. This shows in particular that $f^\mu|_{\mathbb{H}^*} = f^\nu|_{\mathbb{H}^*}$, thereby proving our claim.

Summarizing, we have a well-defined map $\beta : \mathrm{Teich}(X) \to B(\Gamma)$ given by $\beta(\tau) = \varphi^\mu$, where $\mu \in \tau$ is arbitrary. This map is easily seen to be injective (exercise). By lemma A.2.2, the image of β in $B(\Gamma)$ is a bounded set in the Nehari norm. It remains to show that β is an open map. That $\varphi \equiv 0$ is an interior point of the image of β is an immediate consequence of theorem A.2.4. To show the same for other points of the image, one needs a non-trivial theorem of Ahlfors concerning quasi-conformal reflections across quasicircles, and so we omit the proof. The interested reader will find the details in [A1, pp. 131–4]. □

We recall at this point that when X has finite conformal type, say (g, n), then $B(\Gamma) \simeq Q(X)$ is finite dimensional: its complex dimension is $3g - 3 + n$, by theorem A.1.2. In this case, the fact that the map β above is open can be deduced from the topological theorem on the invariance of a domain (see [Ga, p. 105]).

Corollary A.2.6 *The Teichmüller space* $\mathrm{Teich}(X)$ *of a hyperbolic Riemann surface* $X = \mathbb{H}/\Gamma$ *is a complex manifold modeled on the Banach space* $B(\Gamma)$.

Proof This is now clear from theorem A.2.5. □

A.2.3 Teichmüller's theorem

Suppose that we are given two topologically equivalent compact Riemann surfaces X and Y and that we fix a homotopy class of homeomorphisms $X \to Y$ (relative to the ideal boundary of X). By a standard result on smooth surfaces, such a homotopy class always contains diffeomorphisms. In particular, the set of qc-conformal homeomorphisms in this homotopy class is non-empty. The compactness principle for quasiconformal maps shows that in this set there exists a qc-homeomorphism having the smallest possible maximal dilatation. Is it unique? The answer is yes, and this is a special case of the following deep theorem due to Teichmüller.

Theorem A.2.7 (Teichmüller) *Let X and Y be two finite-type Riemann surfaces and let $h : X \to Y$ be an orientation-preserving homeomorphism. Then there exists a unique quasiconformal homeomorphism $\phi : X \to Y$ homotopic to h which is extremal in the sense that $K(\phi) \leq K(\psi)$ for every other quasiconformal homeomorphism $\psi : X \to Y$ homotopic to h.*

We will not prove theorem A.2.7 here. For a complete proof, based on the so-called Reich–Strebel inequality, see [Ga, chapter 6].

In fact, it can be proved that the extremal mapping ϕ in the above statement either is conformal or has a complex dilatation given by

$$\mu_\phi(z) = k \frac{|\varphi(z)|}{\varphi(z)},$$

where $0 < k < 1$ and φ is a holomorphic quadratic differential on X (possibly with poles at the punctures of X) such that $\iint |\varphi| \, dx dy < \infty$. A mapping ϕ with these properties is called a *Teichmüller map*. Geometrically, a Teichmüller map is (essentially) an affine stretching in the coordinates on X provided by the quadratic differential φ. In this sense, Teichmüller's theorem can be viewed as a strong generalization of Grötzsch's theorem.

The hypothesis that X and Y are of finite type cannot be removed, as shown by Strebel.

A.2.4 Royden's theorems

H. Royden proved two beautiful theorems about Teichmüller spaces which admit a complex structure, such as Teich(X) for an X that is

a surface of finite conformal type. When Teich(X) has a complex structure, there are at least two natural metrics there, Teichmüller's and
Kobayashi's. The general definition of Kobayashi's metric on an arbitrary complex manifold is completely analogous to the notion of the
Kobayashi (pseudo-) distance on a ball in a complex Banach space, given
in section 5.1. Royden's first theorem states that these two natural metrics in Teichmüller space are the same.

Theorem A.2.8 *Let X be a Riemann surface whose Teichmüller space
Teich(X) admits a complex structure. Then the Teichmüller metric
agrees with the Kobayashi metric on* Teich(X).

We will not prove this theorem here. See [Ga, chapter 7] for a complete
proof.

The second theorem of Royden characterizes the *isometries* of Teichmüller space. If $h : X \to X$ represents an element of the modular
group Mod(X) then h induces a self-map $T^h :$ Teich(X) $\to$ Teich(X),
as follows. If $[\mu] \in$ Teich(X) is an element of Teichmüller space (μ being
the Beltrami coefficient of a qc-homeomorphism $f^\mu : X \to X^\mu$), let
$T^h([\mu]) = [\mu^h]$, where μ^h is the Beltrami coefficient of $f^\mu \circ h$. The self-
map T^h turns out to be holomorphic and is easily seen to be an isometry
of Teich(X). It is also clearly invertible, with $(T^h)^{-1} = T^{h^{-1}}$. Thus
Mod(X) acts on Teich(X) as a group of bi-holomorphic isometries. This
group is called *Teichmüller's modular group*. Royden's second theorem
says that this group is the full group of isometries of Teich(X), in all
but a few exceptional cases.

Theorem A.2.9 *If X has conformal type (g, n) with genus $g > 2$ then
the Teichmüller modular group of X is the full group of isometries of
Teich(X).*

See [Ga, chapter 9] for a proof of this theorem and for a list of the
exceptional cases with $g \leq 2$.

Exercises

A.1   Find Aut(X) when X is the annulus $X = \{z : 1 < |z| < R\}$.

A.2   Let $\varphi : \mathbb{D} \to \mathbb{C}$ represent a holomorphic quadratic differential in
       the unit disk. Suppose that $g^*\varphi = \varphi$ for all $g \in$ Aut($\mathbb{D}$) (here
       $g^*\varphi$ is the pull-back of φ by g as a quadratic differential). Show
       that φ vanishes identically.

A.3   Let $V \subseteq \widehat{\mathbb{C}}$ be a simply connected domain such that $\mathrm{Aut}(V)$ contains only Möbius transformations. Show that V is either the whole Riemann sphere, the Riemann sphere minus a point or a round disk. (*Hint* If $\widehat{\mathbb{C}} \setminus V$ has more than one point, let $\phi : \mathbb{D} \to V$ be a Riemann map and use exercise A.2 to show that $S\phi = 0$.)

A.4   Verify that the Teichmüller distance d_T defined in subsection A.2.1 is indeed a metric in $\mathrm{Teich}(X)$.

A.5   Show that $(\mathrm{Teich}(X), d_T)$ is a *complete* metric space. (*Hint* Use the compactness property of quasiconformal mappings.)

A.6   Show that the Nehari norm is complete.

A.7   Let $F : \widehat{\mathbb{C}} \to \widehat{\mathbb{C}}$ be a local homeomorphism. Show that F is a *global* homeomorphism *onto* $\widehat{\mathbb{C}}$. This fact was used in the proof of the Ahlfors–Weill theorem A.2.4.

A.8   Prove the following fact, also used in the proof of theorem A.2.4. Let μ_n be a sequence of Beltrami coefficients with $\|\mu_n\|_\infty \leq k < 1$, and suppose that $\mu_n(z)$ converges to $\mu(z)$ pointwise almost everywhere. If f_n and f denote the normalized solutions to the Beltrami equation for μ_n and μ respectively, then f_n converges to f uniformly on compact subsets of $\widehat{\mathbb{C}}$.

A.9   Prove that the map β constructed in the proof of theorem A.2.5 is injective as claimed.

References

[A1] L. Ahlfors. *Lectures on Quasi-Conformal maps.* Van Nostrand, 1966.

[A2] L. Ahlfors. *Complex analysis.* McGraw-Hill, 1953.

[An] M. Andersson. *Topics in Complex Analysis.* Springer-Verlag, 1997.

[AB] L. Ahlfors and L. Bers. Riemann mapping theorem for variable metrics. *Ann. Math.* **72** (1960), 385–404.

[As] K. Astala. Area distortion of quasi-conformal maps. *Acta Math.* **173** (1994), 37–60.

[Av] A. Ávila. Thesis, IMPA, 2001.

[B1] A. Beardon. *A Primer of Riemann Surfaces.* London Math. Soc. Lecture Notes Series **78**, 1984.

[B2] A. Beardon. Iteration of rational functions. Springer-Verlag, 1991.

[BA] A. Beurling and L. Ahlfors. The boundary correspondence under quasiconformal maps. *Acta Math.* **96** (1956), 125–142.

[BC] M. Benedicks and L. Carleson. On iterations of $1 - ax^2$ on (-1,1). *Ann. Math.* **122** (1985), 1–25.

[Be] L. Bers. Finite dimensional Teichmüller spaces and generalizations. *Bulletin Amer. Math. Soc.* **5** (1981), 131–172.

[Bi] L Bieberbach. *Conformal Mapping.* Chelsea, 1953.

[BKS] T. Bedford, M. Keane and C. Series. *Ergodic Theory, Symbolic Dynamics and Hyperbolic Spaces.* Oxford University Press, 1991.

[BL] A. M. Blokh and M. Yu. Lyubich. Non-existence of wandering intervals and structure of topological attractors of one dimensional dynamical systems II. *Ergod. Th. and Dynam. Sys.* **9** (1989), 751–758.

[Bo] R. Bowen. Hausdorff dimension of quasi-circles. *Publ. Math. IHES* **50** (1978), 11–25.

[BR] L. Bers and H. L. Royden. Holomorphic families of injections. *Acta Math.* **157** (1986), 259–286.

[BuC] X. Buff and A. Cheritat. Ensembles de Julia quadratiques de mesure de Lebesgue strictement positive. *C. R. Acad. Sci. Paris* **341** (11), 669–674 (2005).

[CG] L. Carleson and T. Gamelin. *Complex Dynamics.* Springer-Verlag, 1993.

[Da] G. David. Solutions de l'équation de Beltrami avec $\|\mu\|_\infty = 1$. *Ann. Acad. Sci. Fenn. Ser. A* **13** (1988), 25–70.

185

[dF1]   E. de Faria. On conformal distortion and Sullivan's sector theorem. *Proc. Amer. Math. Soc.* **126** (1998), 67–74.

[dF2]   E. de Faria. A priori bounds for C^2 homeomorphisms of the circle. *Resenhas IME-USP* **1** (1994), 487–493.

[dF3]   E. de Faria. Asymptotic rigidity of scaling ratios for critical circle mappings. *Ergod. Th. α Dynam. Sys.* **19** (1999), 995–1035.

[dFGH] E. de Faria, F. Gardiner and W. Harvey. Thompson's group as a Teichmüller mapping class group. *Cont. Math.* **355** (2004), 165–185.

[dFM1] E. de Faria and W. de Melo. Rigidity of critical circle mappings II. *J. Eur. Math. Soc.* **1** (1999), 339–392.

[dFM2] E. de Faria W. de Melo. Rigidity of critical circle mappings II. *J. Amer. Math. Soc.* **13** (2000), 343–370.

[dFMP] E. de Faria, W. de Melo and A. Pinto. Global hyperbolicity of renormalization for C^r unimodal mappings. *Ann. Math.* **164** (2006), 731–824.

[DE]   A. Douady and C. J. Earle. Conformally natural extension of homeomorphisms of the circle. *Acta Math.* **157** (1996), 23–48.

[Den]   A. Denjoy. Sur les courbes définies par les équations differentielles à la surface du tore. *J. Math. Pure Appl.* **11**, série 9 (1932), 333–375.

[DH1]   A. Douady and J. H. Hubbard. Itération des polynômes quadratiques complexes, *C. R. Acad. Sci. Paris* **294** (1982).

[DH2]   A. Douady and J. H. Hubbard. On the dynamics of polynomial-like maps. *Ann. Sc. Éc. Norm. Sup.* **18** (1985), 287–343.

[Dou]   A. Douady. Le théoreme de intégrabilité des structures presque complexes. In: *The Mandelbrot Set, Theme and Variations.* London Math. Soc. Lecture Notes Series **274**, 307–324, Cambridge University Press, 2000.

[Fal]   K. Falconer. *Fractal Geometry: Mathematical Foundations and Applications,* second edition. John Wiley and Sons, 2003.

[Far]   O. J. Farrell. On approximation to a mapping function by polynomials. *Am. J. Math.* **54** (1932), 571–578.

[Fat]   P. Fatou. Sur les équations fonctionnelles. *Bull. Soc. Math. France* **47** (1919).

[FK]   H. Farkas and I. Kra. *Riemann Surfaces.* Springer-Verlag, 1980.

[Ga]   F. Gardiner. *Teichmüller Theory and Quadratic Differentials.* Wiley Interscience, 1987.

[GH]   P. Griffiths and J. Harris. *Principles of Algebraic Geometry.* Wiley Interscience, 1978.

[GS1]   J. Graczyk and G. Swiatek. Induced expansion for quadratic polynomials. *Ann. Sci. Éc. Norm. Sup.* **29** (1996), 399–482.

[GS2]   J. Graczyk and G. Swiatek. Generic hyperbolicity in the logistic family. *Ann. Math.* **146** (1997), 1–52.

[GSS]   J. Graczyk, D. Sands and G. Swiatek. Schwarzian derivative in unimodal dynamics. *C. R. Acad. Sci. Paris* **332**(4) (2001), 329–332.

[Gu]   J. Guckenheimer. Sensitive dependence to initial conditions for one-dimensional maps. *Comm. Math. Phys.* **70** (1979), 133–160.

[Ha]   P. Haissinsky. Chirurgie parabolique. *C. R. Acad. Sci. Paris.* **327** (1998), 195–198.

[He]   M. Herman. Sur la conjugaison différentiable des diffeomorphisms du cercle à des rotations. *Publ. Math. IHES* **49** (1979), 5–233.

[HK] J. Heinonen and P. Koskela. Definitions of quasi-conformality, *Invent. Math.* **120** (1995), 61–79.

[Hu] J. H. Hubbard. Local connectivity of Julia sets and bifurcation loci: three theorems of J.-C. Yoccoz. In: *Proc. Topological Methods in Modern Mathematics, A Symposium in Honor of John Milnor's 60th Birthday,* Publish or Perish, 1993.

[IT] Y. Imayoshi and M. Taniguchi. *An Introduction to Teichmüller Spaces.* Springer-Verlag, 1992.

[Ju] G. Julia. Mémoire sur l'itération des fonctions rationnelles. *J. Math. Pure Appl.* **8** (1918), 47–245.

[KH] A. Katok and B. Hasselblatt. *An Introduction to the Modern Theory of Dynamical Systems.* Encyclopedia of Mathematics and its Applications **54**, Cambridge University Press, 1995.

[Ko1] O. S. Kozlovsky. Structural stability in one-dimensional dynamics. Thesis, 1998.

[Ko2] O. S. Kozlovsky. Getting rid of the negative Schwarzian derivative condition. *Ann. Math.* **152** (2000), 743–762.

[KT] K. Khanin and A. Teplinsky. Herman's theory revisited. `arXiv: 0707.0075`.

[L1] M. Lyubich. Some typical properties of the dynamics of rational maps. *Russ. Math. Surveys* **38** (1983), 154–155.

[L2] M. Lyubich. On the Lebesgue measure of the Julia set of a quadratic polynomial. Preprint IMS at Stony Brook, no. 1991/10.

[L3] M. Lyubich. Combinatorics, geometry and attractors of quasi-quadratic maps. *Ann. Math* **140** (1994), 347–404. A remark on parabolic towers. Manuscript, 1999.

[L4] M. Lyubich. Dynamics of quadratic polynomials, I-II. *Acta Math.* **178** (1997), 185–297.

[L5] M. Lyubich. Dynamics of quadratic polynomials, III. Parapuzzle and SBR measure. *Astérisque* **261** (2000), 173–200.

[L6] M. Lyubich. Feigenbaum–Coullet–Tresser universality and Milnor's hairiness conjecture. *Ann. Math.* **149** (1999), 319–420.

[L7] M. Lyubich. Almost any real quadratic map is either regular or stochastic. *Ann. Math.* **156** (2002), 1–78.

[L8] M. Yu. Lyubich. Non-existence of wandering intervals and structure of topological attractors of one dimensional dynamical systems I. The case of negative Schwarzian derivative. *Ergod. Th. and Dynam. Sys.* **9** (1989), 737–750.

[La] S. Lang. *Introduction to Complex Hyperbolic Spaces.* Springer-Verlag, 1987.

[LS] G. Levin and S. van Strien. Local connectivity of Julia sets of real polynomials, *Ann. Math.* **147** (1998), 471–541.

[LV] O. Lehto and K. Virtanen. *Quasiconformal Mappings in the Plane.* Springer-Verlag, 1973.

[LY] M. Lyubich and M. Yampolsky. Dynamics of quadratic polynomials: complex bounds for real maps. *Ann. Inst. Fourier.* **47** (1997), 1219–1255.

[Man] R. Mañé. On the instability of Herman rings. *Invent. Math.* **81** (1985), 459–471.

[Mar] A. I. Markushevich. *Theory of Functions of a Complex Variable,* vol. III. Prentice-Hall, 1967.

[Mas] W. S. Massey. *Algebraic Topology: An Introduction.* Harcourt, Brace and World (1967).

[McM1] C. McMullen. *Complex Dynamics and Renormalization.* Annals of Math. Studies **142**, Princeton University Press, 1994.

[McM2] C. McMullen. *Renormalization and 3-Manifolds which Fiber over the Circle.* Annals of Math. Studies vol. **135**, Princeton University Press, 1996.

[McM3] C. McMullen. The Mandelbrot set is universal. In: *The Mandelbrot Set, Theme and Variations.* London Math. Soc. Lecture Notes Series **274**, 1–17, Cambridge University Press, 2000.

[McM4] C. McMullen, Riemann surfaces, dynamics and geometry. Course notes, Harvard University, 1998.

[McM5] C. McMullen. Area and Hausdorff dimension of Julia sets of entire functions. *Trans. Amer. Math. Soc.* **300** (1987), 329–342.

[McS] C. McMullen and D. Sullivan. Quasiconformal homeomorphisms and dynamics III. The Teichmüller space of a holomorphic dynamcial system. *Adv. Math.* **135**, 1998, pp. 351–395.

[Mi1] J. Milnor. Dynamics in one complex variable: introductory lectures. Friedr. Vieweg and Sohn, 1999.

[Mi2] J. Milnor. Local connectivity of Julia sets: expository lectures. In: *The Mandelbrot Set, Theme and Variations,* London Math. Soc. Lecture Notes Series **274**, 67–116, Cambridge University Press, 2000.

[MMS] M. Martens, W. de Melo and S. van Strien. Julia–Fatou–Sullivan theory for real one dimensional dynamics. *Acta Math.* **168** (1992), 273–318.

[MNTU] S. Morosawa, Y. Nishimura, M. Taniguchi and T. Ueda. Holomorphic dynamics. Cambridge University Press, 2000.

[MS] W. de Melo and S. van Strien. *One-dimensional Dynamics.* Springer-Verlag, 1993.

[MSS] R. Mañé, P. Sad and D. Sullivan. On the dynamics of rational maps. *Ann. Scient. Ec. Norm. Sup.* **16** (1983), 193–217.

[MT] J. Milnor and W. Thurston. On iterated maps of the interval. In: *Dynamical Systems, Proc. U. Md. 1986–87, ed. J. Alexander,* Lecture Notes Math. **1342** (1988), 465–563.

[MvS] W. de Melo and S. van Strien. A structure theorem in one-dimensional dynamics. *Ann. Math.* **129**, (1989) 519–546.

[NPT] S. Newhouse, J. Palis and F. Takens. Bifurcation and stability of families of diffeomorphisms. *Publ. Math. IHES* **57** (1983), 5–71.

[Pa] J. Palis. A global view of dynamics and a conjecture of the denseness of finitude of attractors. *Astérisque* **261** (2000), 335–348.

[PZ] C. Petersen and S. Zakeri. On the Julia set of a typical quadratic polynomial with a Siegel disk. SUNY at Stony Brook IMS preprint 2000/6.

[Re] M. Rees. Positive measure sets of ergodic rational maps. *Ann. Sci. Éc. Norm. Sup.* **19** (1986), 383–407.

[Ro] P. Roesch. Holomorphic motions and puzzles (following M. Shishikura). In: *The Mandelbrot set, Theme and Variations.* London Math. Soc. Lecture Notes Ser. **274**, 117–131, Cambridge Univerity Press, 2000.

[Rud] W. Rudin. *Real and Complex Analysis,* second edition. McGraw-Hill, 1974.

[Rue] D. Ruelle. Repellers for real analytic maps. *Ergod. Th. and Dynam. Sys.* **2** (1982), 99–107.

[Sh]   W. Shen. On the metric properties of multimodal interval maps and C^2 density of Axiom A. *Invent. Math.* **156** (2004), 301–403.

[Sh1] M. Shishikura. On the quasiconformal surgery of rational functions. *Ann. Sci. Éc. Norm. Sup. (4)*, **20**, no. 1, (1987), 1–29.

[Sh2] M. Shishikura. The Hausdorff dimension of the boundary of the Mandelbrot set and Julia sets. *Ann. Math.* **147** (1998), no. 2, 225–267.

[Sh3] M. Shishikura. Bifurcations of parabolic fixed points. In: *The Mandelbrot Set, Theme and Variations*. London Math. Soc. Lecture Notes Ser. **274**, 325–363, Cambridge University Press, 2000.

[Sie] C. Siegel. Iteration of analytic functions. *Ann. Math.* **43** (1942), 607–612.

[Sin] D. Singer. Stable orbits and bifurcations of maps of the interval. *SIAM J. Appl. Math.* **35** (1978), 260–267.

[Sl]   Z. Slodkowski. Holomorphic motions and polynomial hulls. *Proc. Amer. Math. Soc.* **111** (1991), 347–355.

[ST]   D. Sullivan and W. Thurston. Extending holomorphic motions. *Acta Math.* **157** (1986), 243–257.

[Su]   D. Sullivan. Quasiconformal homeomorphisms and dynamics I: Solution of Fatou–Julia problem on wandering domains. *Ann. Math.* **122** (1985), 401–418.

[SV]   S. van Strien and E. Vargas. Real bounds, ergodicity and negative Schwarzian for multimodal maps. *J. Amer. Math. Soc.* **17**, no. 4 (2004) 749–782. Erratum: *J. Amer. Math. Soc.* **20**, no. 1, (2007), 267–268.

[Sw]   G. Swiatek. Rational rotation numbers for maps of the circle. *Commun. Math. Phys.* **119** (1988), 109–128.

[Wa]   P. Walters. *An Introduction to Ergodic Theory*. Springer-Verlag, 1982.

[Yo]   J.-C. Yoccoz. Il n'y a pas de contre-example de Denjoy analytique. *C. R. Acad. Sci. Paris* **298**, série I (1984), 141–144.

[Zi]   M. Zinsmeister. *Thermodynamic Formalism and Holomorphic Dynamical Systems*. SMF/AMS Texts and Monographs vol. **2**, American Mathematical Society and Société Mathématique de France, 2000.

Index